SEX!

(for females only)

BARRY WREN AM FRSN
MD, MBBS, MHPEd., FRANZCOG, FRCOG

To order additional copies of this book, contact:
Xlibris
AU TFN: 1 800 844 927 (Toll Free inside Australia)
AU Local: 02 8310 8187 (+61 2 8310 8187 from outside Australia)
www.xlibris.com.au
Orders@Xlibris.com.au

ISBN:	Softcover	979-8-3694-9591-9
	Hardcover	979-8-3694-9611-4
	EBook	979-8-3694-9590-2

Library of Congress Control Number: 2024908098

Print information available on the last page

Rev. date: 04/29/2024

Introduction

So you think you know all about sex!

One dictionary definition of the word *sex* is that "sex is the sum of the anatomical and physiological differences or the phenomena by which the male and the female of any living species are distinguished."[1] Others suggest the word applies to that attraction that draws one member of a living species toward another. Another definition suggests that it applies to the actual physical act of reproductive intercourse between members of living species, while others apply *sex* to the enjoyment experienced when imagination of a fictitious sexual event increases desire.

Whichever definition of sex you believe is correct, in the real world, sex is regarded as a commercial commodity—like gold. Sex can be exchanged for property or possessions, worn and displayed like an attractive ornament, or bartered during a financial exchange. Sex has many applications, but unlike ancestral dictums, sex is dominated by females. They are now in control!

The delight and desire men have expressed regarding the sexual attraction of the female body is not just an expression of licentious charisma during recent years but has been displayed in sculptures and artistic findings in archaeological explorations of many ancient historical sites. The sexual attraction of females was appreciated a thousand years ago in much the same manner as is expressed today, but there was a difference! *In ancient times, the sole reason and purpose of a female was to please and care for a male.*

It was because of those early differentiating origins that the beginning of gender inequality became evident. Education, art, military power, and science were the domains of males, while females were confined to the kitchen, the loom, and the bedroom. As social civilisation began replacing the hunter/gathering society, working men (such as aristocrats, soldiers, senators, tribunes, and merchants) anticipated and expected to be able to attend a gathering of senior citizens in the local bathhouse in the evening with a glass of wine and some relaxation after a day of work. Although a community bathhouse (for men only) became the centre for vigorous discourse and ingenuity, men were accustomed to being entertained during those exuberant evenings by young females playing music or presenting theatrical or sporting exhibitions. Females were expected to make a man happy! So bathhouses were frequently regarded as centres of sexual enjoyment.[3, 4, 6, 7]

A third-century ceramic mosaic in the floor of the gymnasium in the Roman villa in Piazza Armerina in Sicily. The artwork suggests that as long as two thousand years ago, while at play, young women wore a primitive, bikini-like costume, but quite often, the actual participants in similar games and entertainment may have been completely naked.

In those ancient times, sexual entertainment was presented as a live display so that the audience could enjoy close physical contact with the actors, participate and mingle, and/or actually indulge in sexual activity,[2, 3, 4, 5] whereas in the twenty-first century, sexual entertainment usually alludes to the thoughts an individual has regarding the real or imagined appeal presented in film or on television. Sex in recent years has been associated with imagined fantasy as well as the attraction and the desire that are promulgated by the vision. The audience is often left frustrated and dissatisfied!

Sex is probably the most important coercive enigma in the history of life!

During those incongruous years (between two and twenty thousand years ago), when human life began to emerge from primitive social collectives to eventually develop a sophisticated, thoughtful government that introduced laws and offered education, philosophy, architecture, art, and mathematics to males, females were taught those manners, procedures, and sexual functions that were necessary to please and satisfy a man. While construction, military advances, and power began to materialise in countries such as Greece, Egypt, Mesopotamia, and later in Rome, male entrepreneurs were entering the most interesting and exciting advances in history.[6, 7, 8]

As slaves and pubescent females from the financially destitute caste of society were frequently the most available individuals to provide the entertainment for senators, tribunes, or merchants, the wealthy, educated men found it was more exciting to indulge in some form of sexual interaction with a young slave or a poor, "common" female than return home to his socially acceptable wife (who was probably caring for their children). Female entertainers were required to present physical evidence of their sexual attributes during athletic or theatrical displays, and as a consequence, for many centuries, promiscuity became paramount in most levels of society, with older, wealthy males developing an increased desire for sexual contact with those nubile young women.

In the thousands of years prior to the abolition of slavery, it became a necessity in an impoverished family that any "extra" young females in a large family be sold by their father into slavery rather than search, in a futile effort, for a paid occupation as a free citizen in domestic service.[3, 4, 6] Females were expendable and became a drain on those poor families that had insufficient money to support all members of the family. Apart from her value as a married partner, the only occupation the majority of young women could hope to find, if not sheltered by her father, was as a paid servant or as a slave. As a paid worker, she could anticipate performing menial work in an industrial establishment or housework as a servant to a wealthy household. If

unable to obtain work as a free citizen, "common" adolescent females could apply to a brothel and, if accepted, obtain a reasonable income as a prostitute! Prostitution was regarded as an upper-class occupation for young females in Roman times, but usually only the most attractive, capable young women were accepted to serve in a private brothel. On occasions, when her husband's private brothel was overloaded with clients seeking superior sexual service, an attractive wife in an established aristocratic marriage would be allowed to prostitute herself in order to entertain a man. The art of providing erotic, exciting sexual service was a skill taught in expensive private education institutes, and men were prepared to pay a considerable sum of money to obtain that enjoyment from an educated, well-schooled young woman.

For an adolescent female born into a common, pecunious family, her future revolved around her beauty, her education, her sexuality, and, most important, who would provide or pay for it! Those mostly poor, sometimes destitute, young women were frequently hired as entertainers in the nightly bathhouse meetings, competing with one another and using sexual advances and wiles, hoping to become the most favoured female by one or another of the senior men, and hoping to be selected for a permanent position in his domestic arrangement.

In ancient times, life for slaves and pecunious female adolescents was awful!

Contents

CHAPTER 1

Sex! (For Females Only)

Whether it is a pride of lions, a pod of fish in the sea, birds in the air, or bees in a hive, as well as other living creatures, males for many generations have challenged and fought one another in order to obtain the right to dominate and impregnate females in order to pass on to following generations their particular biological features.

The winner of those cataclysmic physical contests in prides, pods, herds, or collections of wild animals is most frequently the strongest, lustiest young male, who, after vanquishing his older opponent, is allowed to possess, impregnate, and/or cohabit with one or more of the female members of the collection. The right to impregnate females is accepted as the winner's prize following a conflict between males within an animal collection or herd. The dominance of a strong, aggressive male in a collection reflects the desire for that individual to gain leadership as well as the reproductive rights.

None of the female animals ever appear to object to mating with the new dominant young male and appear to readily accept their role of passive sexual partner to the aggressive male! That "prize" is so pervasive throughout the animal kingdom that the motivation is thought to be driven by a particular genetic code in the deoxyribonucleic acid (DNA) of animals.

In studies of the lives of early humans (Homo sapiens), it was evident that physical attributes, including size, strength, and belligerence, of a strong young male led to domination and leadership of collections in much the same way that animals conduct their sexuality in a commune. A desire to become the leader, to dominate, to possess, and to mate with a female was also paramount among primitive humans, and for generations, it led to conflict and physical combat similar to the innate genetic motivation possessed by animals.

So when and why did males develop the thought that they are superior to females? When and how did male dominance and sexual desire become so acceptable that it led to gender inequality in human society?

This narrative recalls and reviews some of the events and consequences that occurred as a result of the myths that led to bigotry and misogyny thousands of years ago, which, unfortunately, are still evident in some communities today.

It is possible, in much the same way that gender domination is present in male animals, that a dominant DNA transcript for moral behaviour may also have persisted

or been inherited through the male DNA in some latter-day humans.[8, 9] However, in spite of suggesting the possibility for a specific mutation (translocation) in DNA for male domination among both animals and humans, the possession of a mutation in DNA of our primitive ancestors has not been identified in animals (or in humans). What has been confirmed, though, is that there are many other segments of the genetic code of humans that have been identified as originating in animals. So it is possible that male dominance may also be inherited.

Human males are generally bigger, stronger, more impulsive, and more energetic than human females, so it was logical that males would dominate those early tribes and collections of our ancestors. Because of brute strength and superior fighting skills, young males were regarded as being more significant and therefore eminent in decision-making and leadership compared to older, weaker men or placid, mild, stoical females! The strongest male demanded that he should lead the tribe. Any objector would be physically beaten in any contest!

And so the most dominant male became the leader not only of his tribe, but also, frequently after combat, he often dominated several collections of other human tribes in order to control a village, a town, a city, or a nation.

To illustrate the significance of DNA in sexual behaviour, it has been hypothesised that between two hundred thousand and five hundred thousand years ago, in an area of Western Europe now occupied by Germany, France, Spain, Italy, and Belgium, a group of primitive humans called Neanderthals lived and survived for centuries.[9, 10] The Neanderthals were just one of several unique species of humans that existed as a result of mutations in their complex molecular DNA. Neanderthals were thought to be larger and probably stronger than present-day humans, but although their DNA sequence was similar, it was not identical to modern humans. It is presumed that some translocated segments of DNA in the Neanderthal species may have produced a different moral code, and it has been hypothesised and suggested that Neanderthals were relatively passive, quiescent, and obedient compared to other members of our tribal ancestors.

Although physically similar to modern humans, it is possible that a particular segment of DNA in Neanderthals mutated, resulting in changes in their social behaviour, obedience, and acquiescence so that aggressive physical defence of their territory, their village, their females, and their family was thought to be unnecessary. If such a mutation occurred, it has been suggested that the general demise of Neanderthals 250,000 years ago was associated with males who rarely opposed those aggressive Homo sapiens who were invading Europe from Africa.[9, 10] It is likely that quiescent Neanderthal females would have been appropriated and impregnated by

the Homo sapiens male invaders (sex was important then—as now). As those females reproduced, the introduction of a different sequence of DNA would ultimately have resulted in diffusion and alteration of the characteristics of both Neanderthals as well as members of other species of humans. Some of those descendants may have remained aggressive and dominant, while others may have retained a peaceful and passive morality. Diverse attitudes and behaviour are present in a percentage of humans in our present population, suggesting that some fragments of Neanderthal DNA may continue to influence physical appearance, behaviour, and morality of present-day Homo sapiens.

What is known of our antecedents is that baboons and humans share over 98 percent of DNA,[9, 10] but about a million years ago, a variation (mutation) in only a few segments of the DNA molecular structure resulted in a huge difference in appearance and outcome of Homo sapiens compared to the physical and mental ability in apes.

It is suggested that over many hundreds of thousands of years, other primitive human ancestors copulated and delivered an infant in which the DNA code is thought to have undergone mutations resulting in a difference in the species, with improved physical and mental capability. The multiple DNA sequences causing particular behaviour in animals (or in humans) have not been identified, but some characteristics of both male and female interactions may be due to inherited chromosomal sequences over which we have no obvious control.

Whatever the underlying reason, and in spite of improved cognitive ability and function, many humans (Homo sapiens) continue to exhibit patterns of behaviour reminiscent of primitive animals. Territorial possession by a human with dominance and possession of their "patch" is regarded as typical of animals as they defend their territory, their sexual partner, their progeny, and their individuality. However, in our present society, some of these attributes are frequently regarded as primitive, selfish, and potentially antisocial compared to the ideal behaviour of a shared, civilised human society.

For thousands of years in many societies, females, physically smaller and weaker, were regarded as inferior to males when considering advisory or leadership roles. As a result, whenever an ambitious, aggressive, or impulsive female demanded an equal voice or elevation to a significant position regarding decisions or leadership, she was met with fierce opposition by males in her community. Because of the subjugation of any ambitious female, the moral and philosophical attitude within a community was consistently driven by the dominance of men.

Until the last few centuries, life for a woman in a male-dominant society was not pleasant. In fact, it was awful! It was only when Pericles introduced democratic

elections in Athens around 400 BC that women gained a voice in political activity, but even then, very few women ever gained any equality. They were destined to take care of the boys!

In Grecian and Roman culture two to three thousand years ago, nubile young women were encouraged to display their sexual attractions by participating in sporting events or entertainment when naked or partly clothed, but they were not allowed or expected to enter politics or to lead their community in planning or making decisions.[2, 3, 4, 5] The purpose of females was to ensure that their man was looked after, sexually satisfied, and happy.

Females who had ambitions for dominant positions or leadership roles were warned that they must keep their thoughts and aspirations to themselves. In those ancient times, those females rash enough to suggest that they had an opinion were often punished by male leaders insisting that in their society, females were only useful for housework, menial tasks, theatrical entertainment, and exotic, lewd, or erotic activity (any such activity in the twenty-first century is regarded as pornographic or libidinous).

From historical remarks extracted from ancient records, the convention in many of those primitive societies suggests that a female was required to obey first her father, then her husband, and finally her son![14] She was discouraged from expressing an opinion regarding her own thoughts or desires. Females, when expressing some objection regarding their subjugation, were frequently deprived of their rights, prevented from expressing their distress regarding their suppression, or punished, expelled, or banished using physical force, including death.

For the greater part of early human existence in the known world, the dominant leader in a collective (tribe) was a strong, aggressive male who, by sheer physical strength, led the action in the tribe. As primitive humans progressed over centuries from a loose gathering of individuals to tribal collections and intimate societies, the urge to possess, mate, and indulge in sex with a particular young female or a certain chosen woman was almost certainly one of the prevailing motivating forces involved in gaining leadership with inclusion of sexual rights in local communes.

Sex with his chosen female was important to the dominant male, who insisted on, and demanded to receive, his prize. Once the aggressive leader was satisfied, other young men regarded other young females as fair game, so sexual intercourse following mutual agreement (or rape if rejected) became a common occurrence in many primitive as well as medieval communities.

For thousands of years, those strong young men had been accepted as leaders, providing protection during conflicts and wars, and they were expected to protect as

well as provide food and shelter for their female companions and their children. Often, the strongest male was so dominant and powerful that following many years as leader, he was regarded as being eternal and given the status of king, emperor, or some other title indicating permanent recognition of leadership. His power usually included the authority to dictate the government of the tribe, town, or nation as well as the right to choose the sexual reproductive rights with one or more suitable females in the tribe who had knowledge and skill in managing home and family. It was helpful if their chosen partner was also young, intelligent, attractive, desirable, and more than willing to participate in sex!

Very few men (or females) in the tribe were physically capable of opposing the choice or decision of a strong dominant male! In those primitive societies, strength and power were equated with intelligence and leadership, so most in the tribe agreed to the demands of the strongman. Physical power and sexual rights were in the ascendency!

In Greek and Roman mythology and in Judaist stories in the Old Testament, strongmen like Samson, Hercules, Goliath, Shadrach, David, Daniel, and Alexander the Great were recorded as being heroes and leaders because of their strength and athletic skills but, unfortunately, not always because of superior intelligence. Very seldom were females given any credit, acclamation, or leadership roles, in spite of being admired by many of their peers and companions. Gender inequality was firmly established in those historic times, and any female who thought otherwise was likely to be punished, rejected, excluded from her social gathering, or even exterminated.

As often occurred, an individual who had been accepted as the most powerful and dominant person, and who had been acclaimed as the leader by his community, eventually became so consumed by self-esteem and ego that he felt that the authority and power he held should pass without dissent to one of his progeny—that one of his male offspring should also inherit that right and should be acknowledged as the next leader of the community.

To achieve his aim and maintain continuity of hereditary leadership, modification to the previously acceptable rules of succession (the strongest, most powerful winner takes all) was necessary. Problems and objections to hereditary succession were frequently raised by some members of the commune (a son might be too young, too weak, or easily defeated by some other male), so it became necessary for those old rules to be overturned or discarded. Overcoming obstacles or objections often resulted in major conflict within the tribe or the family, and it frequently included elimination and death of opponents! Dominant despots, by sheer physical force, imposed their ideas until hereditary power in their country or nation became accepted.

Hereditary power resulted in presidents, emperors, czars, kings, and various other forms of national monarchy being accepted in many nations for hundreds (even thousands) of years. This dominant system of autocratic rule persisted until rebellion by opposition leaders insisted on the introduction of a form of democracy. The first known major success involving opposition to a king or national leader was achieved in 1215 in England when the Barons of the United Kingdom of England led a united opposition to the tyrannical rule of King John and forced his signature to the Magna Carta.

Following that uprising by the aristocratic strongmen of England in 1215, the next thousand years saw democratic elections for leaders being introduced by many European countries and nations around the world. However, in spite of many instances of citizens admiring and endorsing a particular dominant man as a temporary leader, there was always some demagogue (dictator, tyrant, emperor, king, or despotic ruler), bolstered by words of adulation or by imposition of illegal action, who wished to continue familial power and, of necessity, control the thoughts and physical activities of all the citizens in their country.

Fortunately, the elevation of individual, dominant, strong but unelected leaders to perpetual power has been rejected or gradually discarded over the last few hundred years (with the exception of a few totalitarian governments, such as North Korea, Russia, and China). To maintain total control, dominant leaders needed to use the physical power of their armed services to suppress opposition to their dictatorial position, but, as has occurred in many nations, an upheaval eventually occurred, and a form of democracy prevailed.

It is important to note that among all those who sought prolonged authority to lead, subjugate, or dominate a nation, there were seldom females with such an individual desire. Cleopatra of Egypt, Catherine the Great of Russia, and Elizabeth the First of England achieved their adulation and accession of leadership not by individual physical strength or military might but by employing affection, close physical contact, or sexual intimacy to gain advice and support when deciding their affairs of state. Passive, feminine, opportunistic, sexual influence was proving to be more effective and less bloodthirsty than the traumatic damage males had exercised to achieve their goal.

In reviewing the influence of women in politics and leadership over the centuries of recorded history, it is important to consider the role that women have also actively taken in support of their partner. Intelligent and energetic females, using allure and conditional sex, became recognised as a major asset in the role of a supportive partner to any male who held a principal position in the collective.

Perhaps the lesson from history suggests that sex may yet save the world from extinction!

Whatever you think, without the attraction and allure of feminine sexuality, human life would not exist!

CHAPTER 2

The Hierarchy of Sex

For thousands of years, males appeared to dominate the sexual hierarchy in the gender relationship, but during the past few centuries, females have begun to not only achieve equality but potentially may, in the future, dominate the gender partnership, as admiration and acclamation of their beauty, their sexuality, and their prowess prove to be the most significant feature in a relationship.

In those ancient years (between two thousand and ten thousand years ago), including early civilisations in countries such as Egypt, Babylon, Mesopotamia, and early Grecian society, as well as later during the Roman era, life for young females was very tenuous. Males were dominant, and their future was secure, usually in farming, politics, the military, or commerce. But for females, their future was uncertain.

Whether born into wealthy families or into slavery, every female was required to find a position that provided physical protection, food, and a shelter for her future. Adolescent women in wealthy families were given an education, but common families could not afford to educate their daughter, so she was required to support herself as soon as she could gain a placement. The possibilities that existed for young females in ancient societies was by marriage, by entering a protected sexual relationship with a senior older man, or by obtaining a position in elite domestic service. Some political arrangements resulted in the pubescent female becoming the commodity in a financial barter or dowry between two families. Failure to gain a protected relationship frequently resulted in a young woman being forced to seek work in a brothel.

Young females realised early in life that they must develop particular skills and enticements to attract the interest of a male who could provide shelter and enhance their adult lives. Girls were taught, both at school and at home, that sexuality was probably the most important of various juvenile educational manifests. Young females were encouraged to display their body, their potential sexual attributes, and their skills during enticing and alluring artistic displays as they competed for a permanent position in an aristocratic domestic setting or sought to gain the approval of a young man seeking a wife.

Enchanting, pubescent females hoping to capture the attention of a worthy partner were encouraged to display their glamorous and erogenous features during sporting and theatrical activity. For all those centuries, sex with an attractive and popular

female was for sale to the highest bidder. Love was seldom involved in the majority of sexual relationships, but prostitution was a key to financial security for a large section of willing pubescent females in society.

Sex for sale was a dominant theme for hundreds of years in ancient Greek and Roman society as well as during the medieval years in the remainder of Europe.[2,] [3, 4] In Portugal, for many years, convents, nunneries, and cloisters were regarded as little more than royal brothels, with nuns providing sexual service to the king as well as to other members of the aristocratic members of society.[11] According to known records of Portuguese royalty in the sixteenth and seventeenth centuries, King Dom Joao V had several concubine nuns available for sexual antics in various nunneries or monasteries around Lisbon, with his favourite being Sister Teresa da Silva, who was abbess in charge of the St. Dennis monastery.[11] Sired by the king, Sister Teresa delivered two children, one of whom was a son who later became inquisitor general of Portugal and who was responsible for many impositions and lethal interrogation of Protestants and Jewish citizens.

Similar dissolute relationships apparently occurred between nuns, priests, the clergy, and aristocrats in many European nations in the two thousand years, in spite of the introduction of Christianity into Europe. Sexual excesses between both male and female clergy were not only neglected by many Christians, but the pleasure of sexual intercourse was endorsed, enjoyed, and abetted by many nuns and priests, including several papal heads of the Christian religion.[12]

The desire for sexual interaction between males and females has been dominant in Homo sapiens as well as in every animal since creation, and attempts to curb the lust for sex among humans is doomed by natural instinct and ardour, whether or not a religion or cult suggests that abstinence from sex is a pathway to a better future.

In recent times, in spite of efforts to reduce premature sexual activity among immature humans, particularly during the past few centuries, contact between individual male and female adolescents has led to many loving unions as well as many conflicts. The history, as recorded in novels or plays by writers such as Voltaire,[13] Shakespeare, Zola, and Austen, tell of some of the intimate sexual interactions between young men and women who were involved in emotional stories, great love affairs, occasional destructive and/or anguishing political events, heinous crimes, gender criminality, pathos and despair, or scenes of triumph when sexual love and successful emotion prevail.

However, it appears that after a few thousand years of subjugation, gender inequality has begun to change in favour of females! Many women have inherited or developed profitable commercial businesses or occupations and consequently feel financially

secure. Seldom during the last five hundred to a thousand years has any woman been forced by financial deprivation to engage in prostitution, while those women who do enjoy sex for sale continue to willingly entertain clients. Many women, particularly during our most recent century, have exhibited their glamour and sexuality for financial gain and have been known to gain acceptance as respected businesswomen, enjoying the income associated with their sexuality as a commodity for sale.

Advance has been achieved in human knowledge regarding the science of what makes a girl into a beautiful young woman.[16, 23, 27] Understanding the processes involved in the reproductive function of young women is now accepted as a remarkable scientific advance in human physiology. The result is of immense significance and importance, particularly in regard to the efforts involved in maintaining good health in postmenopausal women.[23, 26] Understanding the intricate scientific events, beginning with her conception, her birth, her childhood, her rapidly developing pubescence, and finally into her crowning achievement as a beautiful young woman, has resulted in admiration, respect, and adulation as an adult. While males have frequently demanded their rights in a liaison, it is actually females whose attractive, commanding sexuality possesses the key to feminine control and progress in human sexual behaviour.

This narrative is a review of the effect that sex has had on the social, political, cultural, religious, and cosmetic participation of females in various human collections, tribes, societies, and associations over the past years. The question often asked is "Do you think female relationships with males are different now, compared to those distant years in the past, when it was presumed that the sole purpose of females was to please males?"

To determine the present opinion of females, I asked a group of women to respond with one or two words only when asked what the word *sex* meant to them. The responses were spontaneous and an insight into the attitude of modern females. Sara, a senior company manager, thought that the word *sex* described or alluded to a pleasurable activity. Elema, a technician, was surprised by the question but responded by expressing a joyous "Wow!" when considering what sex meant to her. Nicola, a young receptionist, regarded sex as being a fun experience.

For females in our present society, the word *sex* suggests an exciting, pleasant physical interaction with another human, which may involve intimate, select anatomical portions of her body being displayed or employed to obtain pleasure.

Has anything changed?

Sex remains provocative!

CHAPTER 3

Sex Politics

The indigenous race of Homo sapiens (humans) can be divided into two distinct species, male and female, each of whom develop distinct physical, psychological, mental, and sexual functions that are not present or applicable in the other gender. As a result of those different functional attributes, it is not possible to compare the physical, sexual, or psychological elements in the two separate species. They should be considered separate entities, not compared regarding the purpose or value of contrasting attributes.

An attractive pubescent female born into an aristocratic family a couple of thousand years ago could anticipate that she would eventually become a very important person in the society, but she had to win the plaudits of the ruling family first. For thousands of years during past political eras (Greek, Roman, German, Chinese, Russian, etc.), aristocratic young females were schooled and modelled by parents and special teachers in an effort to gain education and skills while being trained as a potential partner to a dominant male in their society.[2,3,4,5] Any female who contemplated or hoped to attain a superior position in society as the partner of a male ensured that her intellectual skills, as well as her social, physical, and sexual attributes, were displayed. She employed all those features in order to attract the attention, admiration, and desire of the dominant male.

In those ancestral years of emerging human civilisation, the use of and the attractions of sex were mandatory in the educational curriculum. For many years, intelligent, attractive, and charismatic females employed a range of wiles and devices, including their sexual potential and availability, special skills, talents, devices, arts, and knowledge, in order to advertise their advantages to a dominant male of interest.

Once chosen and confirmed as the partner of one of the principals in a commanding position, she was, by subtle manipulation and sexual opportunism, able to exert influence on various major decisions as she persuaded and advised her dominant partner regarding political activity in their ever-changing society.

In the modern era, by employing charismatic influence, sexual expedience, superior skills, knowledge, and other advantages and benefits, females have reversed the primitive animal model of "spoils to the strongest dominant male and as a result have regularly gained a senior position involving control and power in many institutions and decision-making positions.

In some societies in our present world, a number of males still contend that because a physically strong, aggressive man is allowed to dominate his home, his politics, or his nation, he should be recognised as being superior to any female in skills, intelligence, and leadership. That form of gender inequality is now being challenged in different countries, societies, and religions, and the inequality regarding leadership in various communes is the cause of many displays of social unrest and anger.

Because of the difference in moral, philosophical, and sexual response in different species of humans, it is inevitable that there will be different moral and mental outcomes to various provocative stimuli. As a result, neither females nor males should be required to compare the merits of their physical, psychological, political, mental, or sexual attributes. Unfortunately, there are still those who promote those with male strength and bullying as an indication of superiority in leadership activities. Men and women are both members of the human race, but their abilities to command leadership, purpose, and thoughts are as different as racehorses are to farm horses! The majority of societies now accept that males and females are different species of the same Homo sapiens race, and because of the different hormones produced by each species, they both enjoy a variety of individually different skills not present in a member of the opposite gender.

Before reviewing the advances and recognition that females have achieved over the last few hundred years, it is necessary, based on historical records, to reflect on the attitude that males have long employed in order to trivialise and persecute women in society.

The physical harm, the cruel punishment, the derogatory statements, and the social dictates suffered by females over many centuries have gradually been expunged or have disappeared from records, but some societies, religious communities, and nations still retain the basic belief that females are inferior and should be subservient to males. Most advanced communities and democracies now acknowledge that physical violence and derogatory remarks about females are unjust and unworthy of males, and instead, most males admire females as a different and attractive species of Homo sapiens.

Admiration of females by males is based on beauty, intelligence, knowledge, the appeal of youth, and intimate sex following close contact. Humankind has changed, and most males now not only accept females as being different and more than worthy of a place alongside those strong, dominant males, but many men now actively extol the benefit and joy associated with feminine skill, achievement, ideas, thoughts, and virtue. However, even though males in our present society have acknowledged the

worthiness of females in both domestic and community activities, there are still some who hesitate and express doubt about the opinions and advice of a woman.

What was life like for females over those past centuries, and has it changed in our present egalitarian society? Is it possible that some males continue to repeat those same misdemeanours perpetrated during those thousands of years prior to the present era? It seems that some men still regard females with derision and taunt some as being an inferior branch of Homo sapiens.

From written history, life for centuries past was so cruel that women were frequently subjugated and treated like farm animals. Invading armies such as the Vikings as well as the Huns and even the Crusaders[6,7,8] captured whole tracts of land, populations, and citizens and used females and young boys as currency, selling boys, girls, and women as slaves or into prostitution.

Apart from those females who were born into the aristocratic clique of society, the majority of women were subjugated to that coterie of society who were taught that the main purpose of a woman was to satisfy a man. Make him happy!

The purpose and aspirations of a woman is certainly different now compared to what it was a few centuries ago! But has the sexual obsession of males really changed? And is the ambition of women different from those expectations centuries ago? If so, why? How has age, education, science, and understanding influenced attitude and acceptance of changes in society?

CHAPTER 4

Sex Education

It is generally recognised that the majority of adolescent males reach their maturity sometime between sixteen and twenty-five years of age, following which period of maturation they are expected to begin a career, acquire a partner, marry, begin a family, and own their home.

However, in marked contrast, females begin sexual growth and development from about the age of ten to twelve years, and because of early growth of breasts, menstruation, and other obvious gender distinctions, females obtain a form of sexual understanding, sexual morality, and sexual maturity considerably earlier than do males. Many pubescent females have not only developed attractive breasts, feminine hips, and the body of a beautiful young woman by the age of thirteen or fourteen years but also have begun to ovulate and menstruate and potentially could become pregnant. That premature growth and sexual development of females compared to males occurs because of earlier production and release of essential sex hormones in females.

The early pubescence in females often results in young women who have begun exploring their gradually maturing sexual organs experiencing pleasure and excitement. They discover the sensual enjoyment of delicate stroking and feeling their breasts and nipples, their vaginal area, their labia, and their clitoris. So advanced is their physical sexual development and the pleasure they obtain from exploring and enjoying their emerging sexuality that some of those adolescent females express interest in and contemplate the possibility of increased sexual enjoyment and pleasure following insertion of a male penis into their vagina.

Because most males who can consistently gain a penile erection associated with sexual desire are usually some years older than similar sexually advanced adolescent females, it is not uncommon for adolescent females with an experimental desire for sexual intercourse to attract the sexual attention of older males. To achieve that goal, she may have accepted and encouraged the opportunity for sexual intercourse with a mature male. Some young females have not only explored and actively participated in sexual intercourse but may have delivered their first child while still in school. The question arises then, are females in the twenty-first century more promiscuous and sexually aware than in previous centuries?

Almost certainly not!

CHAPTER 5

Sex for Sale

For centuries in early European and Middle Eastern countries, as well as other communities, the discrepancy in familial support for girls in lower-class families was very different compared to that directed to boys.

Males dominated the politics and customs of society for thousands of years. Wars were fought by males, farms were tilled by males, government decisions were made by males, towns and countries were planned and controlled by males, and women had to do whatever their man asked of them.

Having a fit, young, unmarried woman in a family was an expense that few families could afford. She had to bring home enough money to pay for her daily food and care, or she became a liability. If she became a financial burden to the father, it was accepted that she could be sold as a commodity to the highest bidder. Females born into a financially poor family were viewed as a financial problem, and as a result, she had to be sold or otherwise exploited. The use of females after the age of ten years as merchandise was a common trading practice in order to obtain money or reduce the financial strain. It was accepted by multiple societies, from early Chinese, Indian, Middle Eastern, Greek, and Roman cultures until at least the nineteenth century, that too many female children in the family increased the debt or burden on all the members of the family.

Poor parents (for at least the past two to three thousand years) who depended on working sons for support as they aged often sold their extra female children into an organised marriage arrangement or into domestic servitude. Some young girls were sold into prostitution, slavery, or some other form of sexual enterprise because there was no other career available for females.

While a conniving father who trades the sex life and character of his adolescent daughter is condemned with disgust in our present century, it must be viewed in the context of morality and individual circumstances during the few thousand years before, during, and since the demise of the Roman Empire.

A woman's main task was to please her father, her master, or her man!

Discovery of premature sexuality and the exploitation of their sexual appeal was not confined to young daughters in financially destitute families. Aristocratic families also had been trading their young females in order to consolidate family property titles, treaties, and their position in society for many years.

For centuries, it was accepted that when a female developed breasts and/or began to menstruate, she was old enough to begin sexual activity. Many young women were married to older males as part of political manipulation in an attempt to gain political influence or financial benefit. Sometimes the bride had not yet developed any evidence of sexual maturation, with little or no breast development, her labial or pubic tissue not presenting evidence of sexual organ maturation, and her vaginal hymen as yet unruptured. But that did not prevent some impatient groom from trying to penetrate his bride when she was still only a child. The future of such arrangements was full of anger and bitterness or more frequently resolved by harsh physical suppression and beatings. Life for such females was crude and depressing!

The future for common-class females in the past centuries in Greece, Rome, the Middle East, Palestine, Scandinavia, Russia, Turkey, Egypt, Spain, France, and England was bleak unless an attractive young woman could entice some male to buy her as a permanent sexual partner or as a slave. Even then, when she had delivered his children, managed the home and servants, and grown old in his service, she ran the risk of being relegated to the kitchen while her master took a younger female for bedroom duties.

For those young women who were unsuccessful in obtaining a permanent position as a sexual servant or partner to a wealthy male, the future looked bleak. Gaining a position as a prostitute was not easy, for legal brothels required the owner to have good, clean premises, a guard, a bath, and some form of entertainment. A destitute young female, prepared to accommodate a male for sexual service, would find it difficult to accommodate patrons even when she was prepared to offer her sexual services. She could only hope that a wealthy patron would accept her sexual availability if she could provide a superior sexual service to the customer. Some rejected pubescent females were forced to beg, steal, or starve unless conscripted into the most menial of tasks.

Because of our abhorrence of the shameful manner young females were subjected to during the past two or three thousand years, it is relevant to review the domestic attitude by society in those times compared to the life of aristocratic females during the past few years. Catherine the Great of Russia was a fourteen-year-old princess when her mother, the ruling princess of Anhalt-Zerbst in Germany, arranged her marriage to Prince Peter (the heir to the throne of Russia). She was officially married to Peter when she was only seventeen years of age but allegedly had many lovers before, as well as after the assassination of her husband. According to the reputation that Catherine had acquired during the long years of her reign, she had many lovers, and rumours regarding her nymphomaniac desire suggested that she even had two court maidens whose sole purpose was to sexually try out young officers before she accepted their

serviceable sexual ability. Although alleged to be driven by her enjoyment and need for daily sex activity, she became a very important and cunning politician in planning the future development of Russia with the aid of several of her political lovers.

Lucrezia Borgia, the illegitimate daughter of Pope Alexander VI, was married in 1493 at the age of thirteen. She had several lovers both before and after her husband was assassinated. She had several husbands as well as seven children with different men before dying at the age of thirty-nine years. She had been used as an attractive piece of sexual merchandise by her father, the pope, who manipulated her liaisons, marriages, and the deaths of her husbands as he sought to enlarge his hold on both the church and his power in Italy in the fifteenth and sixteenth centuries.

Marie Antoinette of France was an attractive adolescent (an archduchess of Austria) when her mother arranged a political liaison with Louis XVII, emperor of France, and only fourteen years of age when married to Louis. Before being beheaded in 1793, she had enjoyed many sexual partners and had four children, some of whom were acknowledged by King Louis, as well as a black female baby said to have been fathered by a Negro dwarf. Rumour had it that the female child of that liaison was spirited to a Swiss nunnery, where she remained for the remainder of her life. From historical evidence after the French Revolution, Marie Antoinette was recorded as owning a large collection of pornographic drawings and had entertained many lovers at sojourns in her palaces. Sex was foremost in her years as queen of France, and that probably played a role in the decision to end her career on the guillotine.

During the past five hundred years (and almost certainly prior to that time), there have been many similar historical arrangements of pubescent females being married or sold to some other aristocrat or autocrat in order to confirm or solve a financial or a political problem.

Sex with a pubescent female was a valuable commodity for centuries, and young attractive females were treated as an expensive item of commercial trade.

From written evidence and description of events, political exchanges, health records, and monuments commemorating people, disease, and treatment as well as punishment and untimely death, a picture of life for lower-class females over several thousand years can be drawn. Only females born into aristocratic or wealthy families were educated and protected from abuse and sexual persecution.

From early Greek, Roman, Hebrew, and European records and literature, it was recorded that common females were regarded as being expendable and, for a large amount of their life, ranked behind a good horse, a bullock, or a slave when the master considered purchasing household goods. Young males who were considering a career expected to live as a farmworker or to enter military service or eventually become

apprenticed to a successful artisan or businessman. Some handsome young men were able to obtain a position as a servant or slave for one of the many aristocratic females. He was expected to perform at his mistress's soirees like a prized bull, particularly if his mistress demanded his sexual service. For many hundreds of years, until at least the early twentieth century, young females were employed in aristocratic households to be available to assist either males or females for bedroom service whenever guests stayed for a few days of hunting or fishing. Many of those evenings resulted in a servant being punished or discharged if she became pregnant.

To help with the family finances, it was common for young girls or adolescents to be sold by their father to older men to begin life as a sexual partner. While some entered domestic slavery, others were employed as an apprentice in manufacturing or trade. Some became a prostitute (prostitution was considered an acceptable vocation in the Roman commune).

From historical data[2, 3, 4] obtained in records and memoranda from the past few thousand years, it is likely that the majority of females born into families experiencing poverty began having sexual intercourse, following an attachment to older men, within a few years of beginning menstruation. Those who failed to obtain a regular financial stipend were often forced to work as a prostitute.

Immature daughters (often ten to fifteen years of age) of wealthy families were frequently used as the most desirable commodity in barter arrangements to amalgamate the legacy of powerful families and hierarchies. Adolescent girls or immature female daughters were the main commodity available for fathers to provide when bartering with a senior member of an opposing land or country in order to improve his political liaisons or financial gain. If the daughter became pregnant, it strengthened the confirmation of the legal and filial relationship.

While life events for a female from a few hundred to thousands of years ago have often been dismissed or discarded by our present youthful audience as happening long ago and therefore not relevant now, it is important to realise that almost half of the population of our world is still being administered to by politicians who are not particularly interested in the integrity of feminine sexuality or morality. As a result, many incidents of sexual abuse still occur.

In our present "moralistic" and "virtuous" society, the same lewd and erotic sexual activities described in Greek and Roman art and literature of those times are broadcast regularly on our television, on our computers, and in our local theatres. Sex with a beautiful young female is still the most favoured winner for all seasons and for all times!

Not only do men in our wealthy society spend a very large portion of their income, their time, and their objectives in search of sexual gratification, observing young women conducting themselves in exactly the same way that financially destitute young females performed for older citizens in ancient times, but the same entertaining display is now available in our lounges (the modern bathhouse) with a glass of wine as we watch a sexual display presented at house parties or on our television screens. Similar to thousands of years ago, these beautiful young females dress in transparent or scanty clothing, wear thongs or bikinis, and ensure you can see their long, smooth legs as well as a sexually stimulating glimpses of their breasts. Because of the circumstances, they are delighted with the desired sexual effect on the male audience. For a little more applause, appreciation, sexual stimulation, and possibly some financial component, some are willing take all their clothes off before climbing into your bed (exactly the same as the events thousands of years ago in a Roman bathhouse).

Young women from financially deprived sections of our society realise that for fun and pleasure with an older well-healed male, they can gain security and money—at least for a few years. Sexual allure, desire, and enjoyment have not changed over the past two to three thousand years, and neither has society.

CHAPTER 6

Sex—Is It Lawful?

One of the great anomalies in our politicised societies is the punishment of males for a sexual relationship with an adolescent female who has reached sexual maturity but, by age, is still regarded as a child. The legality of sexual intercourse as an offense is currently based on the age of the female when the event occurred, not on the possible maturation of reproductive organs or the young woman's adventurous desire for sexual enjoyment.

The recent history of females in our society, when becoming sexually mature as children, is no different from the situation as existed so many years ago. An American singer, Elvis Presley, began an amorous relationship with the fourteen-year-old daughter (Priscilla) of a colleague in their army. He eventually married her some years later. Another singer, Jerry Lee Lewis, married a schoolgirl (Myra Barton, the daughter of a religious parson) when she was thirteen years of age, and he had sexual activity with her until she obtained a divorce. While there are many instances in Asia, England, and Europe of adolescents (schoolgirls) developing sexual relationships with older males who were delighted with the pleasure at the prospect of having a young virginal schoolgirl as their sexual partner, it is now regarded as illegal in most jurisdictions. Evidence exists that some of those immature females frequently encouraged the sexual adventure, and by accepting the exploitation and opportunistic enterprise, they enjoyed the relationship. In spite of it being considered improper, the majority of males who responded and encouraged exploration of sexual activity by schoolgirls have not been punished.

While complaints by female children are relatively rare, very few clandestine relationships have been investigated or brought to the notice of the legal department. In many instances, the reluctance is due to the fact that the sexual activity may have been instigated and even enjoyed by the adolescent female, who did not wish her sexual exploration and desire to be discovered.

CHAPTER 7

Sex—Historical Mythology

In the Torah (that includes the Hebrew history of those ancient times), it was written that it was Moses who declared that God had chosen a special site as the area for human habitation when making the world. Historical references suggest that the land between two rivers was the site chosen to be that Garden of Eden.

The Tigris and the Euphrates Rivers run out of the mountains in south-eastern Turkey onto the fertile soil once occupied by tribes of Sumerian, Akkadian, Assyrian, Hebrew, and other ancient collectives. Because they fit the bill, the area bound by those rivers is thought to be the site of the Garden of Eden.

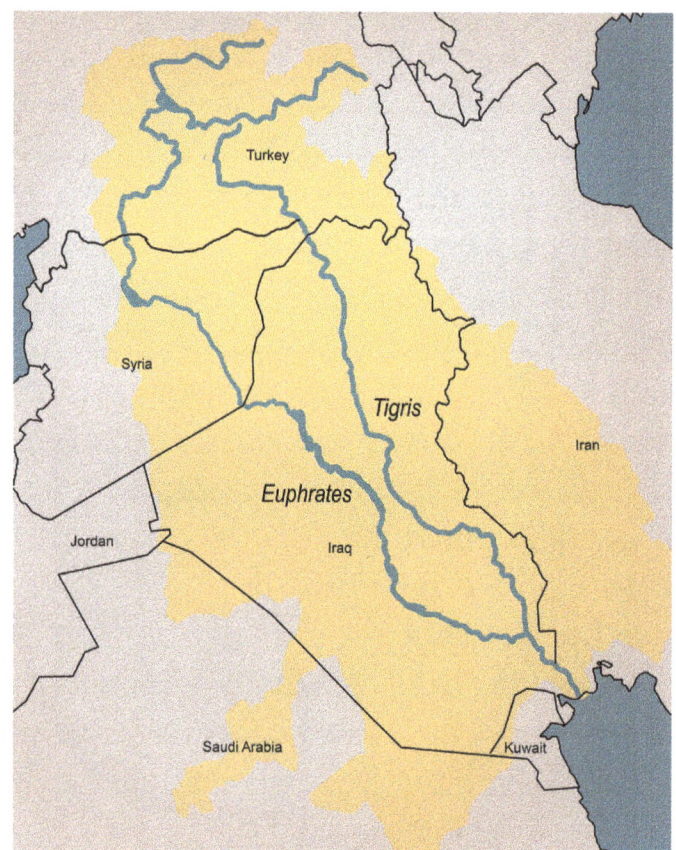

The area bound on either side by the Euphrates and the Tigris Rivers (presently shared between Syria, Iran, and Iraq) was presumed to be the site of the Garden of Eden, but for thousands of years, the inhabitants of the area experienced frequent differences of opinion, expressed by different tribal leaders, as to who owned or controlled various sections of this precious soil.

Following the scripture about the creation of the world in those times, the same God apparently placed all creatures (including plants, animals, fish, birds—and Adam) onto those fertile lands.

It was also written in the Torah that after creating the Garden of Eden, with instructions to Adam to attend to the animals and plants in the garden, Adam apparently felt lonely and asked God to provide him with a companion to help him tend the garden.

So God created Eve. One of the original myths created by religious leaders to explain the genesis of humankind states that Eve was created from one of Adam's ribs, with the express purpose that she was to be a helpmate as well as an intimate (sexual?) companion to Adam—to be fruitful, to multiply, and to replenish the number of humans in the Garden of Eden.

Because the Garden of Eden was new, innocent of pollution and virginal in a primitive state, it must be assumed that Eve was young and naked when Adam first set eyes on her.

The words written in the book of Genesis regarding the creation of man were originally accepted as the true story regarding the genesis of humans. The mythical story suggests that God expected them to lead a virtuous and happy coexistence. Sex would have been preeminent! Unfortunately, however, following some controversy and dissention involving a snake persuading Eve to eat the forbidden fruit from the Tree of Knowledge, Eve was expelled from the Garden of Eden. She was regarded as being a temptation for men and thus the original troublemaker.

The original concoction about the creation of the world and the origin of human existence, as written in the Torah, continued to be accepted for centuries, during which time it also became incorporated as the Old Testament in the early history of Christianity. For many hundreds of years, both preceding as well as following the crucifixion of Christ, the story of Eve was regarded as part of the authentic history of the creation of the world. That fanciful story from the Torah, regarding the expulsion of Eve from the Garden of Eden, continued to influence Western civilisation for many years, especially following a declaration by the Christian evangelist Paul, many years after the crucifixion of Christ, that the head of every family was to be man and that a woman must be respectful and subservient to the male in the family.[17, 18]

Women were regarded as necessary for cohabitation, to help with work, and to reproduce children for the family, but they were never accepted or regarded as equal to men. Females should not only be subservient but also be a willing, fit, and attractive companion. And so, because of such religious mandates, misogyny, bigotry,

and gender inequality were established, and for hundreds (possibly thousands) of years, this fable dominated the role of women in some societies.

In spite of the fable regarding the creation of human life, the religious story of Eve, and her expulsion from the garden, most males continue to believe that all women are beautiful and are a joy to behold, from infancy to maturity.

Young females had always been admired by males, but unfortunately, as they aged, not only did many women cease to maintain their youthful allure, but they also lost their desire for work. And if they also failed to reproduce children, many were frequently replaced by a fertile, younger, nubile female for bedroom duties, while the original partner was relegated to the kitchen. Life for elderly females in ancient societies (and not so ancient) was not a pleasant episode, as many were relegated to a lower status in society, banished, discarded, or even terminated (as being a witch).

As a result of the biblical myth about the genesis of Eve in the Garden of Eden, women have been forced to experience centuries of inequality, suppression, and harm. While primitive ideas circulated in some religions and societies, the negative interpretation of that old fable is not accepted in modern society, so it is important to review the scientific process involved in the magical sequence of events that creates every woman.

Unfortunately, the sexual creation and the scientific processes controlling her unique reproductive function is still a mystery to a large proportion of our modern civilisation. If not created by God, how is she created? (More about that later!)

Apart from the Torah, the history recorded in the early years of human occupation of our world does not comment on the relationship between Adam and Eve, but we must assume the relationship was to be based on sexual desire and reproduction. Adam, like all lonely males, would have been delighted to have Eve as his companion, but because she sinned by eating from the Tree of Knowledge, she had to be punished, and Adam would have been mortified when she was expelled. He could no longer trust a woman! As a result, females must be admonished and punished!

There is no doubt that all women are beautiful, yet instead of observing, enjoying, and applauding women, some men have, for thousands of years, based on advice from religious sources, community leaders, and other elders in the community, attempted to subjugate, segregate, and/or punish females in order to keep them in their place.

So what was the place for women? Why have some men continued to treat women as second-class citizens, slightly more valuable than an animal on a farm?

It is beyond the moral instinct in modern men to regard females in the same way that we currently treat our domestic pets, and for that reason, it is inconceivable that some males, after thousands of years of attraction, admiration, and familial affinity,

still treat some females as if they were a merely part of the housekeeper workforce or a slave. Even though we contend that we live in a progressive society, some men have continued to regard a female as a favoured house pet!

We can ask, "How is it that this obnoxious attitude continues in our society?"

Fortunately, women nowadays do not placidly accept their role only as a sex partner, a housekeeper, or merely an incubator and carer for the next strata of the family. But they do enjoy being lauded as the creator of change in society.

In modern democracies, a young female is acclaimed not because of inheritance of power or familial financial support but because of her skills, intelligence, charisma, and attraction as a beautiful and intelligent woman. By knowing and understanding the science involved in the creation of women, we can now appreciate and enjoy the difference between human females and human males.

However, many religions and cults continue to promote their version of stories and myths that circulate in their closed societies regarding gender inequality. It is important that the source of many of the myths should be explored in order to correct the accuracy of historical or religious tracts that still persist and mislead. The female and the male genders should not be compared as if one was more worthy than the other. Both genders have distinct advantages not present in the other, and credit should be proffered accordingly.

CHAPTER 8

Sex and Mythology

Why is it so?

For thousands of years past, whenever a small child asked their father, "Why is it so?" the father would explain what he knew, but when the child asked a question about a subject the father did not know, the father, rather than admitting his ignorance, often concocted a response that the child forever accepted as the truth.

That fantasy or mythical response was eventually repeated and passed down as the truth (or lore) of the family by the son to his children's children in years to come. This often continued through many generations, to be accepted as the truth.

This chapter reviews some of the ridiculous responses by fathers, village elders, religious leaders, and crackpots who have been quoted regarding the function and health of a female. Some attempt at codifying female function during those past years was developed by philosophers and health medicos (Aristotle, Hippocrates, Galen, and other philosopher physicians) or by physiologists, anatomists, and anthologists who had little medical or scientific knowledge.

Over past centuries, when physiological knowledge and the anatomical differences regarding the function of many feminine sexual organs as well as the unique behaviour of women caused men to wonder what made them tick in a manner so different to males, a number of explanations were promulgated to interpret what they never really understood. Perhaps sexual desire provided a reason for vague explanations or ridiculous answers in ancient times, but truth regarding the physiological cause and effect of feminine individuality lies in the science of procreation.

In ancient times, females had only one purpose in life—to obey and satisfy a man. Provide him with sex, respect, and compliance. The masterful male was supreme, and females should not attempt to achieve any other purpose or level in the family or in society.

There is little evidence of the effect of postmenopause on the activity or social behaviour of older females in those ancient days, but reference to inability to procreate was frequently mentioned as a reason that senior males requested their ageing wives to obtain a younger female (a slave?) for their domestic setup in order to satisfy their sexual requirements.

The most intriguing puzzle for the ancient philosophers (who knew everything!) was the necessity of explaining the processes involved in the growth, development, and

blooming of young females. Chief among the bewildering observations of adolescents was the growth of sexually attractive breasts and vulval pads. The sight of the developing sexual organs in an adolescent female was both enticing and alluring for a male, who often went to great lengths to purchase or appropriate a female of that age for his sexual pleasure. (Remember Helen of Sparta was stolen by Paris of Troy in a story of sexual desire?) Other myths were concocted in those ancient times in order to explain an event that could not be understood or observed. Close inspection of young girls as they began to menstruate was often followed by anatomical examination to determine when she would be available for sexual intercourse. If she had developed breasts and had menstruated, she was considered ready for sexual activity.

Not only did pubescent women grow and develop a different and alluring shape compared to males of the same age, but none of the ancient philosophers could adequately explain the processes involved as young females reached adolescence, began to menstruate, became pregnant, and, after rearing children, eventually reached menopause before entering the autumn and winter years of their life.

To our ancestors and their ancient communities, the regular menstrual bleed at a particular phase of the moon was just as bewildering and incomprehensible as explaining the seasons of the year or the cause for lightning and thunder during adverse weather. To overcome their ignorance, the elders frequently invented gods to explain those events that they could not comprehend (Thor, Zeus, Poseidon, Nemesis, etc.), but they did not understand the cause. Nor did they give the name of a god to explain the reason women had a vaginal bleed every month, or why the regular, bloody vaginal discharge mysteriously failed when women reached a particular age in life.

From records available regarding philosophy, health, and the physiology of ancient civilisations, it is apparent that it was the local philosopher, the reigning tribal chief, or a cleric (the wise men) of the tribe or society who provided answers to the question "Why do women bleed following or during a specific phase of the moon?" With no knowledge of physiology or the hormonal influence regarding the cause of or the purpose of menstruation, the philosophers and wise men conjured many bizarre explanations for the monthly bleed.

Philosopher/physicians such as Hippocrates (460–373 BC) and Aristotle (385–322 BC), in Greek schools of hygiene and medicine, postulated that both men and women produced toxins, moisture, and other harmful substances, and therefore it was important that both males and females get rid of these waste products to avoid ill health. Failure to menstruate meant that a woman retained products that influenced and affected her ability to function.

Hippocrates believed the female body was inherently unstable and relied on the menstrual bleed to remove the four main humors (phlegm, blood, black bile, yellow bile) that resulted from activity involved in maintaining the correct function of the female system.

Aristotle taught that menstruation was the method women employed to remove unwanted toxins from their body and that as women aged, they no longer had the energy, the components, or the capacity to produce a menstrual bleed. He proposed that life depended on heat and moisture and that as women aged, their physical weakness resulted in retention of "wet substances" that eventually induced feelings of heat, sweats, hysteria, and other adverse symptoms, such as sweats and tremors of the head. Aristotle explained that the main reason women stopped menstruating was weakness of their system, and the emotional changes of postmenopause were due to retained toxins. He taught that physical exercise helped in excreting toxins and that if they also developed congestion of the breasts, it implied they would develop madness! (His description of the symptoms experienced by older women is an early description of older, overweight women suffering dementia.)

It was taught that men were able to remove excess amounts of moisture by sweating during heavy exercise programs or by hot baths, but women were required to menstruate to remove those adverse products.[3, 4, 5]

Those Greek physicians/philosophers taught that the uterus (the *hyster*) was intimately connected to the brain and that, as a consequence, when women failed to menstruate, the retained uterine products would directly produce an adverse effect on the brain. Women who did not menstruate would become irritable, irrational, angry, or depressed or display other abnormal mental symptoms. Consequently, they diagnosed such women as suffering from *hysteria* (nothing has changed in 2,500 years).

It was claimed that women naturally had colder bodies, and as a result, they would retain more moisture than men, thus accentuating their defects. It was also postulated that a nosebleed was an alternative method for women ridding themselves of adverse products once they ceased to menstruate.

Galen, the famous Greek anatomist and physician who lived and worked in Rome in the second century AD, taught that men were the perfect human lifeform and that women who menstruated were imperfect human beings. Consequently, women were to be relegated to subjugated roles in society until they reached menopause, at which time elderly women achieved some of the attributes and benefits found in males. It was Galen who postulated that much of women's symptoms and problems arose from

irregular bowel habits. He argued that a well-purged woman suffered from fewer illnesses than one who had irregular toilet activity.

Galen influenced knowledge and treatment of females for over a thousand years after his death, as physicians continued to follow his philosophy that only when they ceased to menstruate did women acquire some of the virtues of a male. Physicians were advised to avoid the foul odours that emanated from menstruating women. As a result, a physical examination was never thought to be advisable when a woman felt unwell during menstruation. This advice was current in most schools of medicine until well into the eighteenth century.

Galen, however, was not the only ancient physician to advise that menstruating women should be avoided and relegated to an inferior status in society. Since biblical times, menstruation had been regarded with repellence, as can be seen in the biblical translation of Leviticus 15, where menstrual bleeding is described in the following terms:

> When a woman has a discharge of blood which is her regular discharge from her body, she shall be in her impurity for seven days, and whoever touches her shall be unclean until evening.

Also in Leviticus 20, the Lord spoke to Moses to emphasise that having sexual intercourse with a woman who was menstruating was a sin:

> If a man should lie with a woman in her sickness and uncover her nakedness; he has discovered her fountain, and she has uncovered the fountain of her blood; both shall be cut off from their people.

During the hundreds of years following decline of the Roman Empire (beginning about AD 400), during which era Christianity began to slowly spread throughout Europe, a large number of priests and Christian converts applied the words of the Bible with harsh and brutal rigidity. Those who succumbed to the "delights of the flesh" during menstruation were prosecuted, and if found guilty, they were often severely punished.

It was proclaimed in many city-states in Europe, for several hundred years from about the tenth century, that any biblical sin associated with sexual activity, particularly during menstruation, was to be punished harshly.

For more than a thousand years, menstruating women were not only regarded as inferior, but in some societies, while menstruating, they were often given the same status as animals. They were excluded from the home and social activity until the bleeding had ceased—no sexual intercourse! In those days, the role of a female was

to give pleasure to a male when he demanded what was his due, so she was often discarded until she was clean! Rape and sexual exploitation of another woman as a substitute for his menstruating partner was regarded as a man's right for over two thousand years, and unhappily, this attitude still exists in some societies in the twenty-first century.

Early education and the interpretation of natural events were taught by philosophers and later by monks (as they were the only ones who could read or write). They taught that as a woman aged and lost vitality and nourishment, she was no longer capable of discharging the necessary blood containing the pestilent material. As a result of the retained toxins, older women developed the changes and symptoms that are now known to be associated with fluctuations or failure of ovarian hormone production.

In the twelfth century, a group of monks in Scotland wrote the Aberdeen Bestiary (one of the oldest surviving manuscripts of that era), containing their explanation of the life, death, and physiology of plants and animals, including the following regarding menstruation:

> The menstrual flow is the superfluous blood of a woman.
>
> It is called menstrua from the cycle of the light of the moon, which regularly brings this flow, for the Greek word for moon is mene;
>
> Menstruation is also called muliebria "womanly business" for woman is the only creature which menstruates.
>
> When they come into contact with menstrual blood, crops do not put forth shoots, wine turns sour, grasses die and trees lose their fruits, iron is corrupted by rust, copper is blackened and if dogs eat it, they become rabid. Asphalt glue, which cannot be melted by fire or dissolved by water, when it is tainted by this blood, disintegrates by itself.

Of all the disputes and angst regarding gender inequality and the suppression of females, most are apparently based on false beliefs created by older men, physicians, or religious leaders. It is essential that all those responses by supposed experts be looked at closely in order to exclude mythical tales and unsubstantiated claims.

With due acknowledgment to the vast array of parents, tribal leaders, and ecclesiastics who have attempted to answer the boy who asks his father, "Why is it so?" there are no bounds to the responses, imagination, and acceptance in a father/child relationship. The imaginative responses were frequently not only accepted, but

after being repeated from one generation to another, they often became permanently incorporated into the lore and finally the law of the primitive tribe.

Menstruation, pregnancy, birth, and menopause are all women's business, yet from the beginning of recorded history, it was men who fabricated the fallacious theories to explain why women underwent the physiological changes that occurred during each of the seasons of her life.

The following explanations depend on the age, thoughts, and knowledge of the parent or person who answered, but in almost all instances, the replies confirm the ignorance of the respondent.

The ancient responses (two to three thousand years ago)—the thoughts, explanations, and physical descriptions recorded by males when answering questions about female function, female sexual provocation, female sexual responses, breastfeeding, menstruation, and menopause—were vastly different from responses to a present-day enquiry with modern medical practitioners.

The fact that, at that time, men had no knowledge of, nor any scientific evidence for, their explanations did not stop ludicrous ideas from being promoted as to why changes occurred to women throughout their lifetime. The two different genders in the human race cannot be compared as if one was better or more meritorious than the other. While scientific study and examination confirm that each gender is a wonderful example of the individuality associated with different DNA, different hormones and different physical components developed within each species of Homo sapiens. Sex and sexual components are the dominant differentiating factor!

The history of life contains many stories, fictions, delusions, and fables invented thousands of years ago by fathers and village elders in order to explain the mystery surrounding the difference, the purpose, and the physiological and emotional emissions of females. In Greece and Rome, philosophers, physicians, and religious leaders hypothesised many ridiculous myths regarding the creation and purpose of women. Many of those fallacious stories were accepted as factual, and because they were endorsed by religious leaders, they prevented females from being regarded as equal to males for many centuries! For thousands of years, it was the acceptance of those fables as religious dictum that resulted in women, particularly older women, being regarded as imperfect individuals.

In those ancient days, physical strength and dominance were regarded as more important than intelligent thought when appointing the head man of a village or tribe. Big, strong, but frequently undesirable or unfit males often became leaders.

Males with a good physique, superior strength, and athletic capabilities were regarded as being perfect, whereas females who had a smaller body, weaker muscular

development, and multiple "peculiar" organs or purposes as well as deficiencies, including menstruation, menopause, hysteria, emotional changes, physical collapse, and atrophic deterioration as they aged, obviously could never be considered for a leadership role!

Originating in many tribes and based on many fatuous narratives, anecdotes, and religious fables, the head man or eminent "wise men" or religious dignitaries of the village continued to affirm the fabricated tales regarding females as truths to explain something that was, and still is, a mystery to many.

Pubescent girls and young women have always been admired and respected for their youth, their beauty, and fr the sexual pleasure they offered or were capable of providing to men. It was because of the desire by older men to have an exciting young woman as his consort that many arguments, conflicts, and wars were fought during past centuries. During those halcyon years, nubile young women were consulted (in bed) by senior leaders and thus were able to exert influential advice regarding the action to be followed. However, when those same women aged and lost their allure, they were frequently denigrated, rejected, or discarded. While young women were admired, praised, and desired, they were frequently abandoned as they aged and lost the sexual glamour and charm of their youth.[13]

As a result, gender inequality, even among the elite of society, appears to have existed not only during a few past centuries but for as long as human life has existed. The bias and discrepancy in function were markedly accentuated in the elderly, with some older women eventually being unjustly treated, rejected, or even classified as witches. Rejection, bitterness, and depression are common terms when describing the characteristics of older women who were (and still are) subjected to abuses, who even now are required to live in isolation in retirement homes or granny flats.

So how do we explain the magic that turns a female child into a beautiful, desirable young woman, and how do we explain the cause of female deterioration as women cease menstruating?

CHAPTER 9

Sex Hormones

What makes a woman tick?

What makes her sexually attractive?

At this stage, a little bit of basic science is necessary to explain the mechanism involved in producing the most beautiful woman.

It has been generally accepted for thousands of years that the brain is the major controlling organ in our body, capable of receiving rapid information conducted by nerves from peripheral sources (our eyes, fingers, limbs, body, ears, taste, skin, etc.) and responding rapidly to the messages by some action. Because it is vitally important for our survival, these messages are transmitted rapidly by nerve fibres that are designed for such a task (just touch a hot stove to appreciate how rapidly the message of pain produces a muscular response to such a stimulus).

However, over the last one or two hundred years, it has been discovered that there is also a second, equally important but slower system, also commanded by our brain, using a different communication pathway that ensures that our organs (heart, stomach, liver, kidneys, ovaries, testicles, skin, muscles, etc.) are maintained and continue to function as required. When we eat food, we don't send a nerve message to our stomach to tell it to initiate the digestive process; that intricate mechanism is set in motion by a cascade of blood-borne chemical messengers called hormones.

Important hormones include estradiol, progesterone, activin, inhibin, testosterone, insulin, thyroid-stimulating hormone (TSH), thyroxin, adrenocorticotrophic hormone, cortisone, anti-diuretic hormone, melanocyte stimulating hormone, bone morphogenic hormone, calcitonin, amylin, leptin, ghrelin, adrenaline, nor-adrenalin, aldosterone, androstenedione, and many more that are being discovered almost daily and investigated in many research departments around the scientific world. This system, using chemical molecules in circulating blood to convey a message in order to activate a cell in a target organ, is known as the *endocrine system*. The endocrine hormone system also has its commander-in-chief (the hypothalamus) located in the brain—a remarkable collection of neurological cells capable of detecting and responding to the most minute changes in the level of chemical molecules circulating in blood that passes through the brain.

It is that remarkable system that regulates the homeostasis of our body.

At the time of her birth, every female child has all the anatomical features found in an adult woman, but both she and all of her sexual body parts are immature. As the child gradually grows over several years, her body, her limbs, her head, and her brain gradually grow so that her body and brain begin to develop and learn many functions necessary for play, education, and physical activity. She is growing, but her pelvic sexual organs, her breasts, and her vagina remain immature. It is not until later (usually between about the eighth to twelfth years of age) that hormone messages to her brain suggest that her body has reached an acceptable stage of maturity. At that stage of her life, the central neurological cells in her brain (the hypothalamus, the pituitary, and other neurological centres) initiate the cascade of hormones that over the next few years turn a child into a young woman.

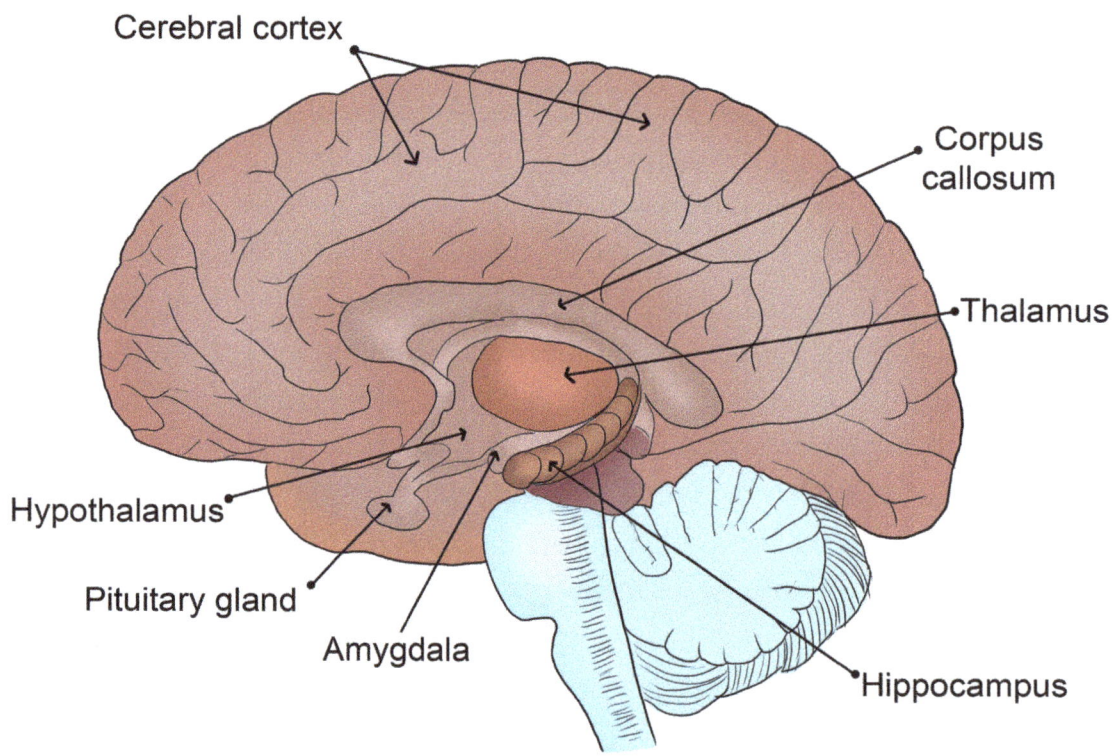

Homeostasis affects total body function. Electrolyte balance, cell replication, organ stability, hunger, thirst, and emotions such as sexual desire, anger, joy, depression, and excitement are just some of the body/brain activities that are influenced and modified by the intricate hormonal system that exists within our body and, depending on the messages received by the brain centres (hypothalamus, etc.), maintains normal homeostasis.

The timing of every task is carefully orchestrated by brain centres, particularly the hypothalamus, in order for each function to take place only when the specific

organ is considered sufficiently mature and physiologically capable of conducting that activity.

The role of hormones (chemical messengers) is to activate cells, maintain normal function and homeostasis in specific cells during growth and development, and to maintain and prolong health in our body. So hormones are not to be disdained, rejected, or blamed. They do not cause cancer; nor are they drugs taken to enhance physical, mental, emotional, or sexual performance. Without the correct circulation of appropriate hormones, we could not grow or survive.

Without the specific hormone, cells will not function. So we must produce those chemical messengers forever to keep our cells functioning. Too much or too little in release of a hormone results in distortion of behaviour from our cells or organs.

The well-being of every man and woman is dependent upon the development of a stable homeostasis from early in life, so the endocrine system (containing probably a hundred or more hormones) is paramount in achieving this state. An imbalance of production of hormones may lead to disease such as diabetes, myxoedema, obesity, menopause, and many more types of dysfunctions, including death.

In young, growing girls, the circulating level of those messages that report data about the physical growth and the stage of the development of many body organs (as well as the activity of her brain) is continually being monitored by the hypothalamus.

The production and release of the many hormones involved in creating a woman are initiated by the hypothalamus as it analyses the complex growth and development of the pubescent female. The cascade of messages includes hormones that stimulate ovarian activation as well as those hormones involved in inducing psychological responses such as enjoyment, emotions, love, sexual desire, and so forth.

The ovaries are two small glands lying on either side of the pelvis. A birth, in a baby, they contain one to two million eggs capable of being fertilised and producing a pregnancy. But until the correct chemical message has been delivered, the ovaries do not release any of the eggs. Not only are most of those eggs never used, but, like hens' eggs, if never used, the eggs deteriorate and expire at a rapid rate. For most women, her ovarian eggs are completely useless or used up around the age of fifty years.

When the baby has grown into a child and the child is growing with a well-balanced homeostasis and is considered mature in physique, the hypothalamus begins the cascade of hormones that are carried in the circulating blood to the pelvis. The special cells surrounding the chosen eggs for that menstrual cycle begin to secrete the hormones involved in promoting sexual cell growth, and the immature girl finds that her sexual organs begin to develop.

Estradiol is the main hormone involved in the creation of that beautiful woman admired by so many males. It is estradiol that is basically responsible for initiating the development of the sexual changes of puberty.

Unfortunately, when all the eggs are used up (usually between the ages of forty-five and fifty-five years) estradiol, progesterone, and other ovarian hormones fail to be produced. All the cells that depended on those hormones to grow and produce our beautiful young woman begin to undergo cell failure and atrophy. Without those hormones, the cells and sexual organs (as well as other organs supported by estradiol—bones, blood vessel, and brain in particular), shrivel and shrink. Menopause has arrived.

CHAPTER 10

Sex—Significance of Hormone Receptors

The mechanism of hormonal activity is very complex but can be simplified as follows: Typical female hormones are estradiol, progesterone, testosterone, oxytocin, activin, and inhibin as well as ghrelin, growth hormone, insulin, thyroxin, follicular-stimulating hormone (FSH), gonadotrophin-releasing hormone (GnRH), and many, many more chemical messengers that provide appropriate feminine body functioning. We don't have to stop and remember the multiple special actions required to metabolise a meal after dinner; it is all done for us by hormones, under the supervision of the hypothalamus. Too much release of a hormone (or too little) produces an imbalance in our body function, leading to major malfunction, such as diabetes (insulin), obesity (ghrelin), a variety of types of ill health, or even death.

When the hypothalamus decides that a particular action is required, it sends a hormone message to the target organ, with the order to take the necessary steps to achieve an appropriate response. All those chemical messengers follow a similar pathway:

1. Hormone messages are specially designed chemicals that will only act on a specific target organ if the target cell has a chemical receptor into which the chemical message can fit exactly in order to relay its message. A human female is born with cells in female organs (breasts, vagina, uterus, vulva, etc.) that each contain the specific *hormone receptor* for that messenger hormone only.
2. For the hormone to respond, it has to dock into the special receptor (like a key that can only turn a lock with an identical receptor lock system).
3. The carrier of most hormone messages is usually the circulating blood that passes to all the cells in her body.
4. The chemical message is only able to deliver its command to the target cells that contain the special chemical *receptor*.
5. When the messenger hormone docks with its identical chemical receptor in the particular cell, it forms a chemical trigger that stimulates the nucleus of the cell to perform the task requested by the hypothalamus.

After the hypothalamus receives data that her body has grown to a sufficient size, all her organs (liver, kidney, lungs, gastrointestinal tract, muscles, bones, skin,

and cardiovascular system) are working well, and she appears to be healthy and comfortable with that development, it begins the cascade of hormones that change an immature female into a beautiful woman. The hypothalamus delivers the first of many hormone messages, the first being gonadotrophic-releasing hormone (GnRH) with instructions to the pituitary gland, only a few millimetres away, that it should produce the follicular-stimulating hormone (FSH).

The FSH chemical messenger travels in blood to all parts of the body, but because the target cells containing the chemical receptor are only present in the ovary, the only response is from the ovary.

The ovarian cells begin responding by producing a gradually increasing amount of a hormone called estradiol (oestradiol). Estradiol is the primary feminine hormone that circulates in blood to reach estrogen target cells in the uterus, vagina, vulva, and breasts in order to stimulate growth and the development of sexual tissue as an adolescent begins to mature into a woman.

While estradiol is the primary hormone to achieve sexual maturation in adolescent females, other hormones are also of major importance in order to maintain normal activity in other cells.

No matter how much estradiol hormone is available in the circulating blood, it will have no effect on any target cell or organ unless the target organ has a *hormone receptor* into which the *hormone message* can dock exactly. Like a key that has been manufactured to fit exactly and turn the mechanism in one lock only, the messenger hormone will fit into and activate the estradiol receptor only.

On fitting exactly into the designated receptor, the fused chemical will then activate the nucleus of the cell to perform the nominated cell activity.

Females are born with the appropriate estradiol hormone receptor in their specific feminine cells (breasts, vagina, labia, uterus, and other female tissue cells), but because the cascade of sexual hormones is not released until she reaches a certain stage of maturity, her sexual apparatus does not begin to develop until her hypothalamus decides she is ready.

Any human who, at birth, does not have the appropriate estradiol hormone receptor is unable to respond to those feminine hormones (estradiol, progesterone) that do not fit exactly into the receptor.

An individual born as a female will have the appropriate hormone receptor in her immature sexual organ cells at birth, and those cells will only respond when the estradiol hormone docks with her hormone receptor.

A male infant born without the estrogen receptor will not achieve any feminine cell change, no matter how much female hormone is administered.

With production and release of estradiol, the miracle involved in adolescence has begun for the female child, with development and maturation of her adult sexual and reproductive organs during her adolescence. The gradually increasing level of estradiol hormone from the ovary stimulates growth and maturation of cells in breasts, pubic tissue, the vulva, the vagina, the uterus, and other sexual organs. Adolescent females begin to experience evidence of ovarian activation and ovulation by increasing breast growth, an aching pelvis from her growing uterus and pelvic support tissue, a slight amount of vaginal discharge, and peculiar sexual sensations as the ovary begins to gradually release increasing levels of estradiol and progesterone. Over a few short years, those hormones change an immature child into an adolescent who begins ovulating and menstruating, and her life begins to change from that time on as she achieves sexual maturation. Within one or two years, she is seen as a beautiful young woman, capable of achieving pregnancy and motherhood.

CHAPTER 11

Sex, Myths, and Creation

Recent scientific studies have not only demolished the fanciful fables of the past but have also provided both men and women with the information that allows them to appreciate the true biology involved in the development of the natural features of femininity and the ability to procreate.

When developing beliefs to explain the genesis of humans, our ancestors resorted to what was, at that time, the only acceptable theory—that a supernatural being must have been present to place all the birds, fish, and animals, including humans, on this earth. That supernatural being was given the status and name of God.

The stories about the genesis of man, according to the early Jewish and the Babylonian versions of the development of humankind, had been recorded as religious actuality for those particular societies. When Jesus Christ deviated from his Jewish origins and initiated his religious beliefs, the myth regarding the creation of man and woman, as recorded in the Torah, continued to be accepted as factual and later incorporated as the Old Testament in the history of early Christianity, in the book known as the Bible.

In that Old Testament, it was written that God made the world and all creatures in it. After some time, he also made Adam. Adam complained that he was lonely, needed help, and wanted a companion, so God took a rib from Adam and from that rib made Eve to keep him company and help him with the work in the Garden of Eden. Similar tales accentuating the physical dominance of males in primitive tribal gatherings had been developing for thousands of years and eventually resulted in most of the early communities accepting the dictum that men were not only superior to females but had been anointed to that position by a supernatural being.

Hebrew historians incorporated some of those ancient historical tales expressed by religious patriarchs such as Moses into the historical events regarding their version of the creation of Earth and all humankind. When Jesus Christ deviated from the teachings of the Jewish priests and initiated the religious faith of Christianity, the history of preceding events, particularly the story of the Garden of Eden as recorded in the Torah, continued to be accepted as factual and was presented in early versions of the Christian Bible. As a result, every woman was forever regarded as representative of the original sinner and therefore culpable and deserving of punishment! Female descendants of Eve were forever expected to be subordinate to men.

For thousands of years, similar stories or warped inventions involving a god or gods with supernatural powers were conceived, narrated, and repeatedly passed on to other members of the tribe, usually by male members in cultures, who were our ancestors. Those fables were not only passed down by word of mouth from generation to generation, but eventually, a number of versions of those interpretations were recorded as religious or cultural dogma.

In the creation fable, God apparently made a woman with the sole intention to be a companion and provide service to a man! To accentuate the dominance of males, a tale was concocted in ancient Greek, Persian, and other Middle Eastern anthology that placed any dissent to masculine superiority as shameful or sacrilegious. It was regarded as justified that any man could punish or beat his partner if she even suggested that she had an equal place in society. The Old Testament, containing the tale that implicated Eve as the culprit in the events involving an apple and a snake in the Garden of Eden, made her life worse.

So for thousands of years, women were branded as the gender at fault and were condemned for the whole of eternity to experience a hard time in a male-dominated society. Even the Christian evangelist Paul in his epistle to the Corinthians[17] confirmed the religious belief of the superiority of males by preaching that "the head of every woman is the man."[1] His epistle in Timothy[18] accentuated the subservient role of women when he preached that "women should adorn themselves in modest apparel and learn in silence with all subjection."

For many centuries, those abstracts, concepts, teachings, and religious proverbs dominated the behaviour, thoughts, and actions of many philosophers, physicians, politicians, and dominant males in various societies as justification to explain the excesses men frequently heaped on women during different eras.

While most of our knowledge about the role of women in ancient civilisations is derived from early Greek, Babylonian, Jewish, and Roman texts,[3, 4, 5] a considerable amount of information is also available in Indian, Chinese, Japanese, and Korean literature about the subservient role of women, particularly older women in Oriental kingdoms. In his historical texts about women in ancient China, Cartwright,[14] a scholar of ancient societies, confirmed that similar conditions existed in most Asian societies, where women were regarded as chattels whose main purpose was to please men. Women in ancient China did not enjoy the status, either social or political, afforded to men. Women were expected to be subordinate, firstly to their father, then to their husband, and finally to their son. Often physically ill treated, socially segregated, and forced to compete with concubines for their husband's affections, a woman's place in Asia thousands of years ago was an unenviable one. Asian women, as well as those living in European and Middle Eastern societies, lived in

a male-dominated culture, forever under the weight of philosophical and religious norms that had been created by men who viewed women as inferior beings.

Although the biblical version of the genesis and creation of humankind is now regarded as a mythical story, it embodies the ideologies of our ancient forebears in multiple societies in diverse nations around the known world, that the main reason for a woman was to be a subservient companion and helper of men. Attractive young women were trophies to be added to a man's collection of assets. Frequently, she had been acquired as a showpiece who was protected and guarded until she aged and lost the allure of youth. After that, she was no longer coveted by other men, often discarded as being old, worn out, and unattractive. Her position in his society was no longer assured, and as her bones crushed and broke, her teeth decayed or fell out, her skin wrinkled and her beauty faded, she was rejected, to be replaced by a younger female.

When comparing adolescent females who had such attractive sexual organs before marriage with those same older females many years later, when they had developed flabby breasts, wrinkled skin, and grey hair and suffered from genital organ and vaginal prolapse, bewildered males asked their ancient physicians and philosophers how it occurred and why it was so.

It was the fragmented and disorderly accumulation of knowledge regarding the earlier reproductive and enjoyable sexual years every woman had experienced that eventually led to major calamities and conflicts as their partner expressed disappointment in the deterioration of the body and appearance of the once beautiful woman in her postmenopausal years.

For thousands of years, young women had been idolised until age eventually caused a dramatic decline in the quality and the attraction of their body. It was thought that as they aged, women became weaker and consequently lost the power to produce menstrual blood. Their wrinkled skin, thin, grey hair, flabby, drooping breasts, loss of athletic strength, bursts of temper, and emotional, hysterical outbursts all followed their weakness and consequent inability to produce menstrual blood. The recommended cosmetic advice was to eat more food. Laying down more body fat would induce larger, fuller breasts and smoother skin than found in thin, scrawny, old witches.

In ancient Rome and Athens, the burden of being an older woman was even more dismal, except for those women who were regarded as belonging to the elite aristocrats of their culture. In those ancient times, society generally regarded females as occupying one of several different roles or obligations within their community.

Young girls born into a family of the ruling class were lauded, educated, and admired as they reached puberty and began to develop the voluptuous, seductive figures that attracted the attention of both young and older men. But those aristocratic

young women were seldom allowed the freedom to develop an affiliation or a loving relationship with young men. Instead, they were educated and taught that their future lay in the expectation that when married (hopefully to a rich and powerful but often much older man), they would manage the household and copulate with their husband sufficiently frequently to deliver a male infant in order to establish and maintain the eminence of the family in that ruling aristocracy. Close physical contact or sexual relations involving aristocratic women who committed an adulterous act outside the rules of the elite society were regarded with extreme disfavour and frequently resulted in banishment or death.

For women born outside the elite ruling circle, life was much less restrictive and, as a result, frequently very amoral. To accommodate the needs of older freemen in Athens or Rome, young women, even though free citizens, were, by circumstances, including financial necessity, often coerced into marrying older men or searching for employment as servants. Others worked in various trades or were conscripted into prostitution. Women were regarded as little more valuable than a horse or a bullock, and as a result, peasants often sold their young daughters into slavery or, by necessity, arranged the marriage of their female child to an older freeman.

There was always an abundance of young women from one or other rank of society, or unqualified girls wishing to learn a trade in specialist occupations but who were able and ready to satisfy the needs of males in the community in order to ensure their immediate financial future. In ancient Rome, about 10–15 percent of young women became sexual partners to an older male or turned to commercial prostitution (it was a relatively acceptable occupation). For a woman who had not been endorsed into the acceptable social strata of married life, sex provided a source of income or security for many years. However, some of those sexual companions, after being exposed to years of adulation and male sexual desire while young and attractive, found that after reaching a particular age, due to the ravages of work or subjugation and beatings, their attraction and position in society had been replaced by desolation and abandonment. Whether caused by the onerous task of managing their household or bringing up their children, or because they eventually ceased to menstruate or had begun to develop signs of physical deterioration, older women were frequently rejected as being sexually unattractive or even repulsive and eventually discarded or left to die in miserable, unhealthy hovels. Inheritance of money or property was not universally practiced or permitted by men in those ancient societies. There was no social welfare for older women once they had outlived their sexual allure or domestic usefulness.[2, 4, 5] A woman who had lost her sexual allure was doomed to a life of degradation.

For many centuries, life for little old ladies was not a happy time!

CHAPTER 12

Sex and Menopause

Fortunately, over many years, the anomalies regarding the value of and role of women in our society have gradually altered, and the gender incongruity has changed. Understanding and admiration of the functions, behaviours, and appearance of females resulted in a very different appreciation of the role of women. When recalling the characteristics, awards, and compliments now bestowed on females we have known during our own lifetime, we are very aware of the remarkable achievements and changes that women have made during the thousands of years since Eve was expelled from the Garden of Eden.

During recent centuries, the altered opinion of women has resulted in increasing respect, admiration, and approval following the achievements she has made as a result of the skills and superior abilities she acquired during the whole of her life.

But has anything really changed over recent years?[8]

Even though men regard their relationship with females in our present society as being cultured, egalitarian, and equal, the dominant attitude expressed by some individuals in a variety of cultures or communities conveys the opposite opinion and attitude. Wife beating, female subjugation, and other gender discrepancies as well as humiliation, sexual assaults, and murder still occur, are allowed, or are ignored in the twenty-first century.

Although modern man is less likely to express such behaviour to their partner, some older women are discarded or rejected, and instead of being eliminated, as occurred in medieval times, they are frequently confined to a back room, a granny flat, or placed in a retirement home to fill in the time until she dies.

Why is this so?

How has gender imbalance been allowed to continue or even take place? Is it because men are wicked or intolerant? Or is it the result of thousands of years of perpetuation of the myth that physical strength equates to superior intelligence and dominance of thought? Or is it because many women, as well as their male partners, have failed to consider the reason for the inevitable deterioration that follows loss of ovarian function?

To understand some of the factors involved, it is necessary to review the history of the development of gender differentiation over the past few thousand years.

More than two and a half thousand years ago, Chinese physicians described adverse changes in older women as due to an "imbalance in the body equilibrium," while

Greek physicians described the psychological changes after menopause as hysteria caused by retention of bloody matter in the uterus. Those beautiful young women who had been admired, loved, and enjoyed during sexual activity were beginning to change as they ceased to menstruate. Women who had often expressed such enjoyment during sexual intercourse began to complain of flashes of heat, insomnia, headaches, loss of desire, depression, and aversion to repeated attempts at sexual fun. Like all women in olden times, sexually active women were caught up by the multiple myths that men had concocted as physicians or religious elders to explain the cause and the changes and deterioration that occurred to elderly women. Not only Greek and Chinese women centuries ago, but even now, modern women in their fifth and sixth decades continue to undergo the same distressing changes, symptoms, and physical deterioration after they cease menstruating. Why is that so?

The answer is found in the story of hormones.

Whether experiencing menstrual irregularity or disordered hormone activity or suffering from fluctuating moods, hot flashes, osteoporotic fractures, heart attacks, dementia, cancer, pelvic organ collapse, or general frailty, the vast majority of women will eventually experience distressing symptoms associated with fluctuating or absent levels of ovarian hormones.

The discovery of the complex sequence of hormones necessary to create a woman, followed by the atrophic deterioration that occurs to every female when the ovaries are finally exhausted of eggs and production of estradiol, is, even now, not fully appreciated or understood, yet is one of the most important factors affecting homeostasis, well-being, and sexuality in females.

While women in ancient times began to deteriorate within a few years of menopause, it was obvious that men of the same age maintained their masculinity and dominant role, being physically and sexually active for many years after their partner had begun to decline during the autumn years of her life. None of the philosophers/physicians during those centuries suggested that the testicles of men might be producing some substance that maintained physical well-being. Not only did they fail to recognise the difference, but none of the ancient physicians ever suggested or extrapolated the idea that females may also have a female substance that dried up or failed to be excreted after menopause. Yet for thousands of years, those intellectuals had observed the changes and deterioration in the health and well-being of every woman after she stopped menstruating, and they thought nothing of it and did nothing! It was only in 1821 that a French physician (Charles Pierre Louis de Gardanne) first coined the term *menopause* to describe that time in a woman's life when she passed from her sexual, reproductive years to the beginning of her senescence.

It was the early French physicians who first associated the cessation of menstruation with the onset of physical, emotional, and mental changes. And it was only in 1941 that Albright[26] confirmed that following menopause, the incidents of debilitating hip and spinal fractures in women increased threefold.

So did her sexual desire also decline? (What has menopause got to do with sexual activity?) Older women remember with nostalgia the exciting pleasure of the sexual encounters they participated in when young. They would like to return to those days, but physical difficulty and loss of mutual desire by their partner inhibits many sexual attempts.

So what has stopped modern women and their partners from entertaining the possibility of maintaining the hormonal situation that produced the most beautiful woman during her adolescence?

CHAPTER 13

Sex—Menopause

For thousands of years, it was implied by the physicians/philosophers in Greece or Rome that the sensation of heat accompanied by sweats, irrational behaviour, outbursts of anger, depression, changes in sexual behaviour, and hysteria were all attributed to retention of and/or subsequent imbalance of the four essential humors (blood, phlegm, yellow bile, and black bile). To release or remove the excessive amounts of those noxious elements that continued to be produced but not discharged (and therefore were still present following menopause), some of the early Greek and Roman physicians, as well as doctors practicing medieval medicine hundreds of years later, promoted and encouraged various forms of treatment to remove or extract these toxins.

In ancient Greece, menopause was regarded as the onset of a mortifying and shameful phase in a woman's life. It heralded that time when a woman became mad, bad, neurotic, and disagreeable. Consequently, a woman who became menopausal was often alienated or ejected from her community because of her hysterical behaviour. Certainly her sexuality was severely affected, and she was frequently punished by her partner.

Hippocrates, Aristotle, and many physicians for over a thousand years after their era asserted that the direct link between the uterus (the hyster) and the brain precipitated the many episodes of abnormal behaviour (hysteria, delusions, neuroses) that women exhibited as they aged and that the problem was caused by abnormal function of the uterus. It was taught that the common symptoms in older women, including confusion, memory loss, anxiety, irritability, anger, and/or insanity, were initiated by and a consequence of the failure of the uterus to produce the regular vaginal bleed that was necessary to discharge the toxins.[3, 4, 5, 8]

Throughout the Middle Ages, few men had any desire to have sexual relations with an older woman who was not only garrulous but had a scrawny body and a dry vagina. Treatment for hundreds (possibly thousands) of years included incising the vagina to produce a bleed, opening surface veins (venesection), cupping, applying leaches, and various other hazardous and distasteful procedures designed to remove blood (and toxins) from a woman's body. Women were prepared to accept any therapy regimen that offered the promise of a voluptuous body and a moist vagina. Being sexually responsive was essential for her survival.

It is little wonder that for many centuries, the sex lives of women were disrupted, handicapped, and influenced by the burden that when young and sexually compliant, menstruation (with a bloody vagina) had made them into inferior beings, only to be followed by obnoxious treatments as they aged following menopause.

The impression that it is only during the past couple of centuries that stories, tales, and anecdotes regarding the peculiar behaviour of postmenopausal women have been recorded is erroneous. For thousands of years, older women, particularly in the autumn and winter of their lives, were subjected to disrespect and humiliation as their charm, body, allure, and sexuality dissipated. Enjoyable sexual activity for older females became a memory of the past or was seldom attempted. It was not pleasant or desirable anymore.

Wives of important citizens in Roman times frequently bought or acquired attractive girls and young women from the local slave market in order to keep their husbands sexually satisfied or to stock the brothel that was often attached to the home. In those ancient days, the winter of an older woman's life was miserable unless she was allowed to remain in the household as a kitchen hand or to care for her husband's new young sex partner. Female slaves became house servants or sex slaves, but often it was found to be more satisfying for the wife of the dominant male to purchase a young virgin who had been sold into a life of sexual slavery by her parents.

Women had one role to play, and that was to satisfy the man of the house. Children of female slaves also became slaves within the household, and although often admired for their beauty and sexual expertise, they seldom achieved any administrative position of power. For women, sex was their most valuable asset. Once their sex was devalued by wear and tear or degraded by age, those older women had nothing to offer to their master or owner. Their life was over!

Over thousands of years since those robust years, in both the Middle Ages and medieval Europe, as older common village women became querulous, repugnant, odious, and/or sexually unattractive, they continued to be rejected, neglected, or ostracised in their own village, town, or city.

Older males, the same age as their menopausal partners, were rarely subjected to derogatory expressions as they aged; nor were they isolated, as were many older women. Men of a comparable age to a postmenopausal woman were usually respected, admired, or venerated in ancient times, and a form of adulation or admiration of older males continued to exist in medieval times, all the way through to our present society.

The testicles of these older men continued to produce testosterone until a very late age. As a result, men generally maintained their physical strength, appearance, sexual activity, and attraction for many years after the allure and pleasure of their female partner had begun to decline.

A large number of women who, when young and attractive, had enjoyed a commanding and envious position in society or in their village were relegated to subjugated roles as they aged—except those few who, being regarded as knowledgeable and experienced, became the fount of stored village wisdom. The skills of those older women in managing pregnancies, treating ills and injuries, and being the source of stored memory and recollection of past events frequently led them into trouble. Although they were often said to be able to forecast when some success or disaster would befall the village or predict an outcome for an individual in an assignment, they were frequently said to be endowed with supernatural powers or in league with the devil. Women who were regarded as having mystical capabilities were at risk from the church, for only priests and bishops were endowed by God to predict the future.[19, 20, 21, 22] (It was either heaven or hell.)

It is important to note that throughout all time, the poems, letters, paintings, and other tributes expressing admiration of females have for the most part been directed in praise of young women, with very little evidence of admiration, attraction, or sexual desire with older women.

The substance of historical events and myths retold by narrators over centuries display a marked bias for heroines to be young, while their adversary is often depicted as being a wicked, malevolent, and cruel older woman (frequently a stepmother, a wicked witch, or a jealous queen). For centuries, heroines who displayed leadership qualities, superior intelligence, or athletic ability were depicted in fables as being young, beautiful, and sexually attractive, while their antagonist was frequently symbolised as being an older woman with malevolent, sinister, or untrustworthy characteristics, often deformed by skin blemishes, bone fragility, and deterioration.

Whether wealthy or poor, older women who had passed menopause suffered from the same symptoms and signs that females thousands of years before had experienced and that continue to cause distress. The medical profession had (until 1930) no therapy available to treat, correct, or reverse the ravages of postmenopausal deterioration, but in spite of their ignorance, those old medical practitioners continued to promote dodgy practices, none of which stopped flashes, sweats, or symptoms and signs of decay.

CHAPTER 14

Sex—Gender Dissonance

From Greek and Roman times through the Middle Ages and medieval years to our present generation, those females who had once been admired for their vivacity, attractive figures, and enjoyment of sexual interaction, with many years of adulation by males, found that they were being avoided or rejected from the festivities that they had once dominated as a pubescent young woman.

They had grown old!

Being an old woman meant little or no more sex. Their role in domesticity was severely compromised, and they often realised that they were unnecessary. They were no longer needed; nor were they wanted.

Greek sculptors, artists in India, ceramic craftsmen in Rome and Sicily, and painters throughout Europe, including many admirers in a variety of cultures, have contributed to the gallery of men who have felt the need to record the beauty and sexual attraction of young women, while none of the artistic treasures remaining from those ancient times depict activity involving older females.

Society in ancient Greece and subsequently in Rome, Pompeii, Herculaneum, and many other cities, during the five hundred years reign of the Roman Empire, had several tiers of social quality, with males (even old men) always regarded as the superior beings. In fact, older females were ignored or hidden from the sight of the master of the home. While aristocratic women of Rome were endowed by inheritance with considerable power and influence, men were always given the authority to rule not only the empire but the home. So when a man wanted sex, he discarded his old partner and acquired a new young female, either as a slave or by a financial arrangement with her father.

So why did this happen to females? Was it something about menstruation?

In reviewing the historical explanations (by males) as to why females menstruate, procreate, and eventually reach menopause, it is clear that none of the early philosophers, physicians, priests, and scientists had any realistic concept. It was just accepted that females had a different pubic and genital anatomy that adequately suited male genitalia when sexually aroused and that menstruation was merely a necessary ritual that females performed to rid themselves of bad toxins. Because menstrual blood carried so many toxins and other harmful substances, it was considered inadvisable as well as bad manners for males to indulge in sexual activity with a woman who had a

bloody vaginal discharge. Growing old was not a common occurrence in olden times, with the average life span for women rarely exceeding forty years, but it is estimated that 25–30 percent of females in an advanced community such as Athens or Rome (2,000–3,000 years ago) did eventually become postmenopausal.[3, 4, 5] Not only did those few postmenopausal women have to contend with religious and ancient beliefs about the defects of their gender, but as they aged, those older women also began to experience an increased risk of bone fractures, heart attacks, stroke, dementia, and social isolation. Older women became a drag on the family and on their community. Their male partner ceased attempts at sexual intercourse and/or frequently sought an attractive, nubile, younger, compliant female.

To make matters worse, the pathological changes to the health of those older women resulted in some being subjected to the most brutal episodes in the history of civilisation. For possibly more than one thousand years, until the beginning of the nineteenth century, the multiple physical, emotional, and sexual altercations resulted not only in the isolation and humiliation of elderly women but eventually contributed to the execution of some of those women, following accusations of being a witch.

As women who had been considered sexually attractive when young became elderly, surviving for years after they had reached menopause, a number began to attract unwanted attention, frequently described as a hag, ugly, repulsive, or being a witch. As their allure and sexual appeal faded, as their facial skin became thin and wrinkled, as they lost teeth or developed a "dowager's hump," as their physical attraction changed, and as their wisdom, and their intelligence were ignored, they became neglected and isolated.

For hundreds of years, older women in villages throughout Europe, the UK, and North America were relegated to menial positions in their society.

From the fifteenth to the eighteenth centuries, many men and women in Europe, particularly in Spain (during the Inquisition), Germany, and France (protesting against the papal edicts from Rome), or those brave citizens who dared to oppose aristocrats or corrupt politicians, were charged with treason, heresy, sorcery, or witchcraft and condemned to death by fire (while tied to a stake or a log), hanging, or drowning.

While burning at the stake was initially introduced to terrify protestors or opponents of religious dogma or governing regimes (remember the story of Joan of Arc in 1431), it also became a popular method in villages to eliminate local irritants, such as vagabonds, crones, or obnoxious or unsavoury neighbours.[20] This horrible and gruesome means of exterminating an "evil person" for treason or a criminal act had been used for several thousand years, beginning in early Babylonian and Jewish

communities, but it reached its culmination in Europe, particularly in Germany and Spain between the fifteenth and eighteenth centuries, when not only criminals but an estimated 60,000–180,000 "witches" suffered from this punishment (the large difference in figures by different authors is because some quoted only known recorded data, while others conjectured unauthorised killings[19, 20, 21]). The cruel methods to punish or eliminate an odd person, a female with Satanic affiliations, or anybody who practiced sorcery were eventually extended in some societies to include any un-Christian act, including adultery and buggery. Although it is believed that less than three thousand witches in the United Kingdom were condemned to be killed by various abhorrent means, it was possible from records to confirm that the vast majority of the witches burnt at the stake in England were female (probably between 70 and 80 percent), and most were in the postmenopausal age group.

Men who challenged the autocracy of the local sheriff or leader of the village commune were often vilified as heretics or warlocks, while women were identified as witches who had placed a curse on an individual or the whole village. By accusing a person of sorcery or witchcraft, it was relatively easy to persuade a group of peasants that a benign or natural event was the work of the nominated "wicked" individual. Victims of such accusations were usually subjected to extreme torture to obtain confessions and then sacrificed by brutal and sadistic forms of fire and/or heated instruments.

Such a method of execution became so frequent that common folk accepted that autocratic decisions leading to such a horrible capital punishment were just and reasonable. Throughout Europe as well as the United Kingdom, witches, sorcerers, warlocks, nonbelievers, protestants, and blasphemers had to be destroyed by the most painful means imaginable, so burning at the stake became a popular form of entertainment.[20, 21]

It was also during the Middle Ages that the adverse influence of menstruation and menopause began to have a major effect on females in society.

In those medieval years, a young menstruating woman was not only required to suffer the vaginal menstrual bleed for one week every month, during which time she was required to tolerate the social, religious, and physical stigma caused by menstruation and the abhorrence associated with menstrual blood, but she was usually isolated or inaccessible to friends, society, and potential partners. Sex was avoided, and many young couples were frustrated by the religious inhibition.

However, after she reached the safety of menopause and no further bleeding or repellent discharge occurred, she began to experience psychological anomalies of irritability, anger, depression, eccentricities, emotional and sexual incongruities, and

physical changes that severely affected her relationships in society, with friends, with her family, and with her husband/lover.

For many women, after menopause, the curse of a monthly bleed was replaced by gradual deterioration of their youthful physical appearance. As their skin became thin and wrinkled and their bone matrix lost calcium, leading to fractures and loss of teeth, they frequently developed memory impairment and dementia. For a woman who had reached menopause, life often became a burden, with many suffering from discrimination, harassment, and persecution.

While some older women merely experienced flashes, feelings of heat, outbursts of aberrant, temperamental behaviour, and depression, others were also afflicted with the dowager's hump, loss of teeth, genital tract problems, including vaginal prolapse, urinary incontinence, and/or a dry vagina. Some of those women progressed to dementia as a result of hormone deficiency, malnutrition, and economic abandonment.

Women who had been adventurous lovers enjoying sexual intercourse during their adolescence and early years were frequently regarded as difficult, querulous, depressed, and unhappy after they had ceased menstruating. Having sexual intercourse had become a chore or was painful and was often avoided. Their husband/lover soon began to look for younger, more attractive mistresses who were capable of fulfilling their sexual needs. The rejected postmenopausal female not only lost her partner but was frequently ejected from the home and her bed. She may have become a vagabond or regarded as a witch.

When some disaster befell a village, the crops failed, or some other adverse event occurred, the villagers complained they had been cursed. Most often, the culprit accused of having applied the curse was some elderly woman in the village who, having been regarded as a vagabond or witch, was unprotected by a family of substance. Others may have been in a stage of early dementia or were too incoherent to defend against accusations of sorcery. Sometimes the accusation of witchcraft was directed by a malevolent neighbour to inflict harm or by a husband in a malicious attempt to eliminate an older, unattractive partner. In 1435, the duke of Bavaria accused his daughter-in-law (Agnes Bernauer), who was a commoner and below his acceptable social level, of witchcraft and sorcery, and while his son was away fighting a war, he had her drowned in the Danube!

It must be recalled that following the fall of the Roman Empire about AD 400, Christianity began to spread in Europe, and its introduction was associated with some of the most tyrannical rulers in history. Competition for the minds and hearts of the European population reached its zenith during the eighth century, following Charlemagne's ascendancy in 772 to become the holy Roman emperor of Western

Europe as he aimed to unify the major states in Europe by conquest and subjugation. Charlemagne used the growing influence of the Christian religion in Rome to expand the French occupation of Western Europe and decreed that all non-Christians were to be eliminated. Those who opposed the words of the Bible were to be charged with heresy, blasphemy, or similar antibiblical indictments, and if found guilty, were to be killed, often by fire. There was an immediate scramble by all who could read to convert to Christianity!

It was following that era that chivalry and romantic protection of women became fashionable for several hundred years. Unfortunately, that courtesy did not extend to the ordinary women of the village. They were still regarded as readily available sexual objects by men in the village.

Although Charlemagne died in 814, his edict (following his elevation by the pope to become emperor of the Holy Roman Empire) became the mandate for most of Europe for the next six hundred years, and the church became triumphant, as every passage of the Bible had to be obeyed implicitly. In Exodus 22:18, it demanded, "thou shalt not suffer a witch to live," and those religious men with their inflexible, resolute, and intense beliefs were intent on obeying that command. Mere women who could predict the outcome of some event or explain the essence of some simple environmental outcome were accused of using magic or witchcraft and thus undermining Christian authority, so bishops demanded that such women be put to death, as they were most likely in communion with Satan.[9]

Once labelled as a witch, it was not unusual to be vilified, charged as a danger, and/or convicted of sorcery or some other misdemeanour. Depending on the local legal authority, such persons were often condemned to death.

And what did they look like?

Remember Shakespeare's description (1606) when Macbeth first confronted the three witches:

So wither'd, and wild in their attire
That look not like the inhabitants of this earth and yet are on't?—
Live you? Or are you aught that man may question?
You seem to understand me by each at once her chappy finger
Laying upon her skinny lips:
You should be women:
And yet your beards forbid me to interpret that you are so.

The burning of witches was banned in England in the latter years of the eighteenth century but still existed in parts of Europe until late in the nineteenth century.

Although witches were prosecuted and, as late as the beginning of the eighteenth century, sometimes condemned to death, the last official death by burning at the stake in the UK was Janet Horne in 1727, who was accused by a neighbour of sorcery.[10]

Janet Horne had been a maid to a wealthy lady in Scotland and had travelled widely with her mistress for many years but became pregnant following a brief sexual encounter and lost her paid position. She delivered a daughter who had several physical deformities to her hands and feet, so Janet and her daughter lived a precarious life without support until they settled in Sutherland, where a neighbour claimed that the deformities in the daughter were a result of a failed attempt by Janet to turn the daughter into a donkey. The trial for sorcery lasted less than one day, and when the sheriff in the county found her guilty, Janet was stripped naked, tarred and feathered, and burnt to death the next day. Janet was described as being old (possibly postmenopausal?) and, according to reports, probably in the early stage of dementia.

The role of physicians in those turbulent times must also be evaluated against the complete lack of understanding and knowledge as to what was happening to those perimenopausal women. While we now understand that the moods and emotional symptoms that occur in postmenopausal women reflect the hormone deficiency that begins during the menopausal transition, none of the physicians in those medieval times appreciated the complexity of their problem. Descriptions indicating the manifestation and symptoms of menopause have been described in many historical descriptions, such as that by Dr. Simon Forman, who was consulted by Sir William Monson's wife in 1597. "She had not had her 'curse' (menses) for many months' and as a result 'she was much subject to melancholy and full of fancies."[5, 23]

The desire to maintain smooth skin, a voluptuous figure, and full breasts in order to maintain sexual attraction appeared to be the main focus of ridicule, and even Benjamin Franklin (the great American founding father) wrote a satirical essay[8] in 1757 about the attempts of older women as they used various devices to maintain their sexual attraction. In one essay, he was extolling his advice on the advantages that a young man should seek in an older mistress:

> The face grows lank and wrinkled; then the neck; then the breast and arm; the lower parts continue to the last, plump as ever; so that, covering all above with a basket⋆ and regarding only what is below the girdle, it is impossible of two women, to know an old from a young woman. And as in the dark all cats are grey, the pleasure of corporeal enjoyment with an old woman is at least equal, and frequently superior, every knack being, by practice, capable of improvement.

⋆A "basket" refers to the pannier device women wore in the eighteenth century to extend the width of their skirt.

Advice such as that was also supported in 1832 by Ralph Waldo Emmerson, the famous American poet who wrote that "the age of a woman doesn't mean a thing. The best tunes are played on an old fiddle."

In 1760, James Boswell[15] (ninth laird of Auchinlek, lawyer, diarist, and biographer), when studying law in London, developed a desire to seduce the attractive young actress, the fair Louisa. In spite of developing an elaborate plan, he wrote about his disappointment when his plan was dashed one week when intimate contact was avoided after he discovered she was menstruating. Some eighteen hundred years after Galen had promoted the concept that having sexual intercourse with a menstruating woman was dangerous, his concept was still being obeyed. To compound Boswell's disappointment

and anger, he discovered that the "virginal" Louisa had transmitted "signor gonnorhea" to him when the seductive sexual act eventually occurred one weekend later!

In China, the symptoms (now attributed to menopause) were regarded as due to imbalance of the body equilibrium, and the first known pharmacopeia (the Shennong Bencao Jing, about two to three centuries BC) described a variety of plant and fungal extracts to maintain and improve the body equilibrium. Chamomile, vitex, sage, panax, ginseng, liquorice, tribulus, motherwort, dandelion, and St. John's Wort were some of the plants used extensively by Chinese and Indian herbalists to treat distressing symptoms in older females. Lingzhi was the fungus of choice, but other plants and fungi were prescribed and are still offered by those practitioners of Chinese Medicine, having been prescribed successfully to women for more than two thousand years. If it has been prescribed for that many years, then it must be good!

During those years in England when advances were being made in maths, architecture, science, and health, the physicians and surgeons who treated patients suffering from the many sexual, psychological, and physical changes following menopause continued to advocate ineffective therapies for women. As a result, most women sought information and treatment from other older women and midwives who openly advertised their services, potions, and healing skills.

So prolific were these older women in attracting perimenopausal females suffering symptoms that the accredited male physicians complained that these older women and midwives were depriving them of their rightful practice and income. In 1511, an act was passed by the English Parliament in order to limit the influence and dominance of these "women who boldly take on themselves great cures and things of great difficulty, in which they use sorcery or witchcraft."

For several hundred years, the conflict that existed between certified physicians and those older, wise midwives who proffered advice to those perimenopausal women continued to be debated by the College of Physicians and members of Parliament.[8, 11]

Many of the physicians evinced their opinion that very few women actually experienced any problems; nor did they complain of menopausal symptoms (probably because women were too embarrassed to openly discuss their symptoms with a male doctor). One of their main arguments was that the midwives (or witches) were promoting potions, ointments, magic, sorcery, and hands-on manipulation (particularly breasts, labia, clitoris, and vagina) in their efforts to treat these postmenopausal women. It was claimed that as a result of these therapies, particularly the physical stimulation when applying ointments, that menopausal women developed an increase in sexual desires and lust—a form of masturbation that the patient enjoyed.

It was argued by physicians and priests that most of those old women advertising their treatment regimens were really witches who were disciples of Satan and whose therapy deliberately increased extremes of sexual desire, sexual urges, or nymphomania. As being a witch, with the use of magic and sorcery to induce sexual "perversions," was regarded as anti-Christian, both the church and the College of Physicians demanded that those midwife "witches" be condemned to death. A large number of those village midwives were consequently drowned, hanged, or burnt at the stake.

John Leake, a respected physician, wrote a series of essays in 1777 on the health of women, promoting the concept of good health during a woman's menstrual years in order to avoid poor health and a loss of vitality in her later years.[5, 23] "Women who have lived temperately and are naturally very healthy, escape without much inconvenience, but I have known some delicate women inclined to hysterics and nervous disorders."

In medieval Europe, while comments about the climacteric changes during perimenopause were hurtful to older females, it was the implication that some older, physically crumbling females were witches and in league with the devil that produced the greatest fear and danger for postmenopausal women. Some were accused of being the instigator of a variety of heinous crimes, such as blasphemy or dealing in magic, sorcery, or witchcraft, for which the punishment was death.[19, 20, 21, 22]

Similar laws had already been introduced in a number of European countries, so it was not surprising that England soon followed suit.[22] In 1542, Henry VIII's Parliament introduced a law in which witchcraft was defined as a felony, a crime punishable by death:[22]

> It was forbidden to use devyces to practice, or cause to be devysed, practice or exercise any invocacons or conjuracons of Sprites, witchecraftes, enchauntements or sorceries to entent to fynde money or treasure or to waste, consume, or destroy any persone in his bodie members, or to pvoke any person to unlawful love, or for any other unlawfull intent or purpose ... or for despite of Cryste, or for lucre or money, dygge up or pull down any Crosse or Crosses or by such Invovacons or sorceries or any of them take upon them to tell or declare where goodes stolen or lost shall become guilty.

The act provided that anyone who should "use, practice, or exercise any Witchcraft, Enchantment, Charm or Sorcery, whereby any person shall happen to be killed or destroyed," was guilty of a felony and was to be put to death without the benefit of clergy.

During Queen Elizabeth I's reign (1558–1603), 157 people were accused of using witchcraft to commit murder or defying the words of Cryste (Christ). Almost half were acquitted, but of the eighty-six who were burned at the stake, only nine were men.[21, 22] Most of the women were midwives or older women who confessed to the charge of healing by "oyntments or embrocations" with incantations. Most of their accusers were physicians, bishops, or priests.

While witchcraft and sorcery were a constant threat for older women in England in the fifteenth, sixteenth, and seventeenth centuries, it was in Europe, beginning in the thirteenth century, that magic or witchcraft became most influential. There was fierce competition between priests, monks, physicians, midwives, folk healers, and diviners to gain the favour of local villagers. Those with the best healing record, using charms, potions, plants, herbs, and other aids, particularly incantations while preparing the mixtures, gained the highest approval rating. Not only did the mixtures and blessings by the priests become a powerful cure for all ills, but their incantations contained words from the Bible, suggesting to the common folk that the mixture was endorsed by God. Because those other purveyors of alternative therapy were being regularly hauled before the ecclesiastic courts and found guilty of treason, blasphemy, sorcery, or being in a pact with the devil (for which the guilty were routinely committed to death), the desire to "cure thy neighbour" lost its moral appeal. So physicians, priests, and monks cornered the market.

Life for little old ladies was never meant to be easy!

Even Voltaire,[7] as recently as 1758, when writing *Candide*, his satirical essay about optimism amid corruption within French aristocracy, could not avoid including Candide's feeling of disgust when, after a separation of many years, he met up with his first love and mistress, the once beautiful Cunegonde:

> Candide, the tender lover, on seeing his beautiful Cunegonde all weather-beaten, her eyes bloodshot, her breasts sunken, her cheeks lined, her arms red and chapped, was seized with horror. As she had aged, Cunegonde was unaware how ugly she had become, no one having told her.[7]

And later, Voltaire describes the once beautiful princess Cunegonde as being shrewish and insufferable, growing uglier by the day. The romantic memory of a beautiful young woman had markedly changed during the prolonged interval between the attraction of his young mistress and seeing her again after many years. Candide was dismayed and disgusted when he saw the effect of age, the environment, and harsh circumstances on his once beautiful princess.

Candide was not the only person who had loved, admired, and lusted after a beautiful pubescent female and then been dismayed and disappointed that the adolescent who he had hoped to make love to eventually aged and became most unattractive.

By the later years of the nineteenth century, women in England had begun to reject the principle of male-only political control, and many brave women formed the suffragette organisation and demanded a vote for women in Parliament. Women were disturbed by the humiliation and suppression that existed throughout the civilised world and rejected the interpretation of biblical commands that confirmed gender inequality and the dictum that the main purpose of a female was to be a sexual partner, a mother to his children—and quiet!

In 1912, Lord Curzon, during the suffragette drive for emancipation, was reputed to be outraged at the thought that women might be allowed to vote or sit in Parliament. "If women are allowed to vote, it means the end of civilisation as we know it!"

Has menopause and the loss of youthful sexual appeal been a disappointment and an emotional disaster for aged women, or is it a relief to escape the desire of men who demanded sexual pleasure from the same female when she was so attractive many years previously?

Even in the twenty-first century, with information regarding the process of reproduction and menstruation, and with modern access to many forms of vaginal hygiene products, hormone therapy, and contraceptives, a large number of women living in insular societies still express a sense of relief that the onset of menopause marks the end of the monthly bleed with the requirement of toilet cloth (the rags). Others express relief that they can no longer become pregnant.

In January 2020, author and journalist Nikki Gemmell,[24] in her personal note about her impending menopause, describes the symptoms succinctly:

> The past few years have seen a strange unsettleness, what is this agitation that's taken over my body, calm, optimism? Its hard to be myself. There's a thieving restlessness, an unevenness, and I was never this. Then there's the sleeplessness. The feeling I'm shifting the side of myself and it's a scattier, looser being coming to the fore ; I need to regain firmness somehow. Find ballast. Blaze again.

And later she adds, "Yet I can't wait for the vast freeing of the menopause. To be free of that monthly drag of menstruation, my life held hostage to the great felling tiredness and the hormonal migraines, still, as the body undergoes its regular, deeply female rhythm. Forty years of it—and what a relief when it will finally be gone."

Over hundreds of years, a young woman's success depended not only on her looks, her intelligence, and her athletic capabilities but most of all on her ability to attract the attention and patronage of a distinguished or prominent male. Her charm and the ability to display her merits, achievements, and potential sexuality were the keys to her success in those male-dominated cultures. Although young women, from puberty to late reproductive years, were admired for their slender, smooth, soft bodies, their full breasts, soft, rounded hips, and firm buttocks, their sexual lure was often insufficient to overcome the harm that emanated from menstrual blood, so most maidens went to great pains to avoid displaying any sensual proclivity while menstruating.

For many years, it was the acceptance of the erroneous hypotheses and religious dictums regarding menstruation that resulted in the employment of ineffective therapy, inappropriate advice, harmful devices, or absurd regimens of therapy to manage adverse menstrual and postmenopausal symptoms.

During past centuries, older women were seldom venerated in the way that postmenopausal females, who were once young, voluptuous, and alluring, continue to attract admirers and followers in our present society. Meryl Streep, Julie Andrews, Jane Fonda, and other beautiful women who were idolised when young are still admired and venerated long after they reached menopause. Is that because of their inherited attributes, modern cosmetics, or their use of hormones?

The transition phase (perimenopause), beginning during menstruating years and lasting for several years after vaginal bleeding has ceased, has for many women been associated with such dramatic symptoms that it has become the focus of books, novels, plays, and films. Films such as *Somethings Gotta Give* with Jack Nicholson and Diane Keaton, *Calendar Girls* with Hellen Mirren, *The Hot Flushes*, and *Hot Flash Havoc* have all made jokes and sarcastic comments about the changes and humiliations that were regarded with amusement by our contemporaries. Among the many authors who have written about the trials and tribulations of the symptoms of perimenopause is Jean Kittson[25] in Australia, who wrote a humorous book, *You're Still Hot to Me*, to explain some of the symptoms that cause anxiety and distress to women during perimenopause. Continuing in her humorous vein, she discussed some of the regimens available to treat women during their distress.

Taking an entirely different point of view, author and journalist Nikki Gemmel,[24] who is approaching menopause, in 2020 wrote, "I'd like it done and dusted as soon as possible, thank you. Don't fear the menopause, don't dread it, its releasing surely. Liberating us from the gaze of men, from the function of Woman as Sexual Object, releasing us to be who we really want to be, perhaps."

In 2001, Jeanie Linders in the USA was motivated by her friend's (and her own) symptoms to write the very successful, whimsical musical comedy *Menopause*, which continues to be produced to world acclaim in theatres around the world. All postmenopausal women attending that musical laugh as they claim recognition of the familiar symptoms of the perimenopausal transition, but only a few appear to be aware of or alarmed by the long-term consequences and changes to their health and sexuality caused by oestradiol hormone deficiency.

One of the marvels of nature is that like flowers that bloom at a particular time of the season to attract bees, nature has also provided a similar system for humans to ensure copulation and reproduction of Homo sapiens. The 'flowering' of females during their adolescence is achieved by the development of a beautiful and sexually attractive body during those years when she is ovulating and capable of becoming pregnant. The development of her sexual appeal is a feature of nature in humans, designed to attract a male in order to copulate and to reproduce another human.

So, the cascade of hormones involved in producing the most enticing sexual appearance in adolescent females is a natural event during the time when her ovarian eggs become available for fertilisation.

It is during those years of ovarian hormonal activity that females achieve their most exhilarating and enjoyable life. They look beautiful, they achieve great feats of physical, mental and emotional action and are admired for what they have accomplished.

Throughout thousands of past years, attractive young women had been admired and idolised by males in their society, but in spite of acknowledging the beauty of those adolescent females, those young women had no freedom or independence. Control over their lives was in the hands of either their father or their owner/husband who dictated their behaviour and lives until they reached their menopause and had lost their allure. Life for those older aged but previously beautiful young women was often miserable, lonely and unhappy.

The hormone cascade had ceased - their menopause had arrived and their life had changed. Their ovaries had ceased producing those essential hormones that had been so essential during their adolescence.

For thousands of years the menopause had been regarded as a precursor to a decline in the sexual delights of a young female. However in our modern life and society, it is possible that many of the desired emotional and sexual features of those young adolescent females can be maintained if those hormones that initiated the unique sexual features are employed. Those hormones are capable of not only maintaining cell

activity in sexual organs but are helpful in reducing the risk to the integrity of bone, the cardiovascular system and the brain.

In contradistinction to ageing females in those ancient times, males even in advanced age have continued to lust after beautiful maidens. If permitted, they are physically able to enjoy and give sexual pleasure to younger women during sexual activity – they are still capable of impregnating a woman until they are of advanced age. However, many post-menopausal women in the same age category as their elderly male partner often avoid engaging in sexual intercourse because they have lost the emotional desire, the physical ability or the vaginal and clitoral sensitivity to participate. Such an unwelcome failure for a previously beautiful, sexually active young female is to be regretted because, for women entering menopause, the early use of appropriate sex hormones could have maintained her body, her genital organs, and her emotional desires in the similar way that testosterone from a male testicle maintains his sexual physical health and enthusiasm.

It is important to remember that after almost one hundred years of scientific research and clinical investigation, the use of appropriate hormone therapy not only reduces distressing physical decline but is capable of maintaining the normal function of many important and necessary sexual organs in the human female body. And in spite of many factitious claims regarding the use of biologically natural hormones it does not cause cancer[23,28]. Ageing of cells and organs and eventual death cannot be prevented, but the previously rapid decline after the menopause can be reduced and enjoyable sexuality may continue into old age.

Sex is here to stay! It is a remarkable and enjoyable element of nature that is entirely in the hands of females and because of its universal influence, is an agency that gives females greater power in our society than the brute strength that males once wielded to gain authority.

Go for it girls!

References

1. *The Concise Macquarie Dictionary* (2008).
2. L. Dean-Jones, *Women's Bodies in Classical Greek Science* (Clarendon Press, 1996).
3. R. Fleming, *Medicine and the Making of Roman Women. Gender, Nature and Authority from Celsus to Galen* (Oxford University Press, 2000).
4. Bagnall R et al 'The Encyclopaedia of Ancient History' 2012 ; Wiley-Blackwell
5. Foxcroft, Louise 'Hot Flushes, Cold Science'. 2009 ; Granta Books
6. Frankopan P 'The New Silk Roads' Bloomsbury Publishing 2018
7. Harari YN 'Homo Deus: A brief history of tomorrow'. Penguin Random House 2016
8. Cregan-Reid V. 'Climate Change'. Cassel 2018
9. Bohme M, Braun R, Breier F 'Ancient Bones' Scribe Publications 2020
10. Wragg-Sykes R 'Kindred Neanderthal'. Bloomsbury Sigma2020
11. Molecky M. 'This Gulf of Fire'. Deckle Edge 2015
12. Hibbert C. 'The Borgias and their enemies' Harcourt 2008
13. Voltaire (Francois-Marie Arouet – 1694 -1778) 'Candide' ; Penguin Classics
14. Cartwright M. 'Women in ancient China' Ancient History Encyclopedia 2009.
15. Bain J 'The journals of James Boswell: 1762-1765' Yale University Press 1994
16. Abreu AP, Kaiser UB 'Pubertal development and regulation' Lancet Diabetes Endocrinology, 2016;254-264
17. The Bible – Paul's Message to the Corinthians
18. The Bible - Paul's message in Timothy 2:11-15
19. Wikipedia 2020: Witches in Europe
20. Wikipedia 2020: Burning at the Stake in Medieval Europe
21. Wikipedia - List of People Executed for Witchcraft
22. Collins D. 'Witchcraft Acts'. The Classical Quarterly; 2001
23. Wren BG, Stephenson-Meere M 'Menopause: Change, Choice and HRT' 20013, Rockpool Publishing
24. Gemmell N 'Ringing the Changes' The Weekend Australian Magazine, 25 Jan. 2020
25. Kittson J. 'You're Still Hot for Me' 2014 ; Pan Macmillan
26. Albright F, Smith PH, Richardson AM. 'Postmenopausal Osteoporosis: Its Clinical Features' JAMA, 1941; 116: 2465 -7
27. Bidlingmaier F, Butenandt G, Knorr H 'Plasma gonadotrophins and estrogens in girls with idiopathic precocious puberty'. Pediatric Res. 1977; 91-94
28. Hanahan D. Weinberg R. 'The Hallmarks of Cancer', Cell, 2000; 100: